THE PARALLEL WORLD

GRAVITATION LIFE EVOLUTION

PANTELEY BAHCHEVANOV

ISBN: 1-4196-7023-9

ISBN-13: 978-1419670237

Visit www.createspace.com to order additional copies.

Contents

MEDIA AND PHYSICAL FIELDS

When talk about distortion of the space in the Universe and gravitation waves we are unable to admit this would be possible in a space not filled by some real substance which to transfer these deformations and vibrations. The presence of occasional or permanent streams of material particles not everywhere and in all directions does not response to that requirement.

This is a basic argument of the Hypothesis subject of the current material to claim the space unidirectional everywhere is being filled by continuous substantial media.

To the problems of distortion and gravitation waves in the space follows within the train of reasoning to add the phenomena of mass producing and annihilation. It is known, two identical particle and antiparticle respectively by defined conditions join in to one and disappear, annihilate. The opposite process is one of mas producing when the particles appear and separate in to particle and antiparticle.

We are not able to accept these processes as beginning or finishing from, respectively, to the nothing. It is logical to consider, there, where they disappear, the matter to be saved and the media from where they appear to be a substantial one, but with other a form of existence. This is one more argument to support the hypothesis.

The Hypothesis claims, the Media is filled by primary substance from which the matter of our inert world is being originated. We are unable to detect the primary substance but this does not mean it does not exist.

Introducing the Media that consists of primary substance we make for the world of the physical fields which without that media could not exist. It is a basic factor for our existence and allows explaining a number of still not revealed problems such as the gravitation and the physical fields of charged material particles – the electromagnetic fields.

The Hypothesis considers the existence of three basic types of electromagnetic fields according their source and parameters. Both of them are being realized by the particles of the inert matter. These two kinds are respectively natural and artificial electromagnetic fields. They have limited parameters what is caused by their origin from matter particles possessing mass and inertness respectively. The most particular restriction for these fields to spread is the speed of light.

The third kind of physical fields is created in non-inert media and does not have the restrictions of the inert matter. The parameters of these fields are unattainable for our world and therefore we are unable to detect them. They could exist in a form of hyper fast streams of coded information and as hyper complex integrated coded structures without which the origin and development of life is unthinkable and which are living organisms themselves.

This is the Parallel World in to which there no mystique is and which is strictly obeyed to the regulations and laws of the physics. When we ask the question where are being created these fields we could imagine the cores of the giants celestial objects in the Universe where the status of plasma might suppose to be at the edge between the primary substance of the Media and secondary matter. When put a question Who created the complex coded fields we go to the only answer possible: The Supreme Reason of the Universe. So we have the chance to understand The Living God of Powers not as being a mystical hypothetical dominant but one which existence power and abilities we are able to expound using the physics. Such interpretation we need in our technologically developed world. The said does not influence or affect the traditional religions in the way people to respect God and to pray through the canons and rites of the traditional religions.

GRAVITTION

Physicists have tried for hundreds of years to define whether light is a stream of particles or a set of waves. These waves carry energy from place to place. In the 1860s and 1870s Scottish physicist James Clerk Maxwell formulated a theory that linked electricity and magnetism, and light, to waves of electromagnetic energy. His theory predicted that waves of varying electric and magnetic fields travel through space in the form of electromagnetic waves. His theory was thus strong evidence that light is carried by waves.

Scientists of the late 19th and early 20th centuries believed that the ether was the medium, or substance, that allowed light to travel through space.

They reasoned that outer space must be filled with an invisible medium, without mass, undetectable by normal chemical and physical means, even though it permeated all matter and all space.

Albert A. Michelson, American experimental physicist, formulated an experiment in the 1870s to detect the ether by studying its effect on light. Michelson repeated this experiment more accurately in 1887 with American chemist Edward W. Morley. They attempted to measure Earth's speed with respect to the ether, the hypothetical substance thought to fill empty space. Their apparatus splits a beam of light so that half went straight ahead and half went sideways (Fig 1). If ether existed, they theorized, it should exert more of a drag on one of the beams than on the other as Earth moves through the ether along its orbit. They found no evidence of a difference in speed, however, which led to the demise of the ether theory.

The theory of relativity of Albert Einstein showed that light did not need a medium through which to travel, so belief in the existence of the ether was abandoned.

Why the remarkable experiment of Albert Michelson could not detect ether's wind? It was performed many times during the year, it was perfectly sensible to give the most precision result. There are two basic possibilities.

The first is – ether does not exist, according the statement of Albert Einstein and the second – ether exists, but its features do not allow to be detected by means of material facility.

In the year 1930 the British theoretical physicist Paul Dirac has worked out a quantum – mechanical equation related to the electron, from which result remarkable effects. One of these effects is the existing of anti-particle of electron – the positron, besides a very strange, absurd behavior has been expected for it. Consequently, Dirac has predicted, both the particles annihilate (disappear) by closed contact, emitting energy as a gamma quantum. This "unbelievable" theory has been confirmed experimentally in 1932, when positron has been detected in the space rays.

BASES OF THE HYPOTHESIS

The further material is called "The Hypothesis". It is based on the claim, the elusive substance ether exists. It fills the space uniformly everywhere in all directions.

The hypothesis calls the ethereal space "Universal Energy and Information Conducting Media" (UEICM), besides, the term in the further text is simply "ether".

Ether is not only the medium which allows light to travel through space. It is the basic substance everywhere (base of the bases), from which the matter of the space objects has been generated. Ether is the missed link, without which the physical fields - electromagnetic and gravity ones and the bases of live could not to be expounded. It is the common link that unites the four basic types of interactions – gravitation, electromagnetic, weak and strong interactions.

The Hypothesis follows a line of abstract and deductive considering, strongly and impartially based on physical laws and regulations. That is the only way about a physical category being invisible, inexplicable and denied by the official science. Abstract and deductive considering in that case means to locate the area where to find the true is mostly probable. When Hypothesis contradicts to officially accepted claims, it is searching for logically based and carefully considered arguments.

The Hypothesis concerns absolute respect regarding the Quantum Theory, and mathematical apparatus that describes all the processes, subject of it.

The Hypothesis asserts there has to be a substance which fills everything everywhere within the space. We know very well the structure of our material world that is sensible, visible, detectable, and able to be explored. Further within the hypothesis it would be called simply matter to distinguish from the substance of ether. The substance of ether would consider as being basic or primary. We pay attention and to the antimatter as well but only as far as it is produced from ether by mass producing.

Further, we are interested basically in the substance ether and its hypothetical features, the Universal Energy and Information Conducting Media, and the matter of our world and the space objects.

EXPANDED AND CONDENSED MATTER

For the purpose of the explanation, the particles which build both the Media substance and the matter are simply represented as spheres, as shown at figure 2, A, B.

Fig. 2 represents the relation between the substance of the Media or ether and the matter, and respectively antimatter.

We might say everything everywhere is matter, because of its continuity within the space. But inert matter takes about 4% of the space and considering antimatter as being mirror to the matter it takes 4% more. The rest 92% apparently is not correct to call matter too.

The first reason is: this is an elusive substance which existence is officially denied. The second reason concerns this hypothesis where the elusive substance is accepted as being real and of importance. In that case it is apparently there is essential difference between the matter and antimatter and the elusive substance of the Media.

The next point developing this concept is to find out the hypothetical properties of the substance and its relation to the matter. It is apparently, the object of this is closed approaching a closed to real picture of what is within the space and the origin of the matter and antimatter.

In that way we distinguish ether as being a substance which fills the space everywhere unidirectional and inert matter and antimatter taking their part of 4 + 4 % of the space.

There are two basic forms of existence which correspond respectively to the substance and the matter. The first exists in expanded form and the second – in condensed. The meaning of this assertion is that the matter is condensed form of the substance of the Media.

The expanded form is basic and concerns the substance ether which fills the space uniformly everywhere in all directions.

The condensed form is the matter which is produced from ethereal substance by concentration of mass besides the dimension of matter particle is many times reduced compared to the large particle of ether. The process is illustrated at Fig. 2.

One large ethereal particle condenses producing one matter and one antimatter particles. The process is available only by a jump from the one state to the other. This determines the explosive character of the processes of mass producing and annihilation and the Big Bang.

The transition of mass from one to other form is obeyed to the law of saving the masses. This means, the mass of the large ethereal particle to be equal to the masses of both the produced matter and antimatter particles. Because of the large volume of the particle of ethereal substance its density is slight.

The process of transition is convertible, accompanied by corresponded energy transfers. The basic process is this of mass producing besides a portion of energy is applied to the ethereal particle in a form of gamma quantum. The meaning of that portion of energy is to move the process allowing mass producing and condensation. In principle the large ethereal particle is not only large but and motionless. It contents both matter and antimatter particles in latent form.

The meaning of this explanation is of importance by defining the properties of the elusive Media or ether. The substance ether which fills the Media consists of large motionless particles. Respectively the density of its particles is negligibly small. This defines one more property of the ethereal substance of great importance. This is its great penetration ability. Its great penetration ability means, all space objects, everything within the material world is being immersed into ethereal substance.

The hypothesis is based on abstract considering besides it explains phenomena in principle. For example, the particles of matter produced from ether might be electron and positron, proton and antiproton, but their concrete kind is not an object of our attention.

Considering further the mass producing process we return once again to the condensation as basic transition which defines the properties of the new born matter particles.

The condensation runs explosively, by a jump from one to the other state but in two distinguished sub processes one of which is elusive and thus undetectable. These sub processes by mass producing concern the forming of the matter particle into two separated subdivisions.

The hypothesis puts the emphasis on this considering the sub processes and the result from the transition in that particular way as being of great importance. The problem concerns the continuity of substance and matter. In the concrete transition it concerns the internal micro and nano spaces between the matter particles and sub particles, the Van der Vaal's forces. We consider the particles and sub particles of matter do not interact by empty spaces between them.

By condensation of one large motionless ethereal particle one material and one anti material particles are being born. We leave the particle of antimatter and continue to pay attention to the one of matter.

The process of condensation of the substance by applying of the corresponded portion of energy leads to creation of particle which consists of nucleus of fully condensed substance covered by a sphere of semi condensed substance. The cover of semi condensed substance is elusive thus undetectable but of crucial importance by forming of particles and macro bodies. It is crucial for forming of gravity field and probably of any physical one.

The behavior of the particles is determined by the conditions in which they are and in addition, and the aim to status of balance and lowest energy.

As shown at Fig. 3, a particle of ethereal substance by influence of the portion of energy transfers into two antipode particles of condensed matter by process of mass producing. Both the antipodes are independent.

The particles which build the substance of ether are electrically neutral, have zero impulse of motion and thus no energy. The masses of a particle of expanded ethereal substance and respectively, the derived from it particle of condensed are equal. But, their volumes are quite different. The volume of the particles of ether expands to very large size, compared to the condensed. Respectively, the density relatively to the volume of that particle is very slight.

It might be considered variants about how the ether is structured. One is, all the particles, known, and unknown to have particular expanded version. Other is – the particles, expanded to be uniform, and by various conditions to get their condensed versions. Further in the material would not distinguish the particles of the ether considering them as being uniform ones.

Within the substance of ether expanded particles are in an internal balance. By the process of mass producing they leave their status of balance and become unable to live long if do not find a new balance formula.

The particles of the condensed substance – plasma or cooled matter find their status of balance by means of combine between opposite charges ones. This principle is base of building of our material world. Of course, neutral particles are as well involved in the structures.

Having no energy and impulse of motion, and extremely slight relative mass, the particles of ether determine their features. They penetrate free through the particles of matter, besides the resistance is so low that is impossible to be detected by any facility made of matter, build by condensed particles. This is the reason for the negative results obtained by Michelson's experiment to detect some "ether's wind". The substance of ether is passive and not resistant at all. It could be activated only by the covers of the matter particles or some energy source conducting that energy. This activation is what we call and consider as being "physical field".

Ether is a universal energy and information conducting media. For our concepts, the process of conducting is going practically without inertia.

The velocity of the light is the highest available in the material world. It is limited by velocity of local oscillations of electrons, respectively – their frequency. The electrons, as any material particle that build the matter have the property to be inert. Inertia limits the acceleration of the impulse emitting particle, electron, from there, its frequency of oscillation and velocity of impulse emitting.

If somewhere other kind of mater exists, where, the frequency is much higher; this matter hypothetically could emit impulses of energy of much higher speed, respectively frequency. Ether is Media able to conduct any wave, regardless of its frequency, shape and speed. It is undetectable from position of our material world because of the limit of the speed of light. So, ether conducts as a passive media energy and information from other sources organized in a different way than of our world of inert matter. This concerns the bases of the physical fields and life.

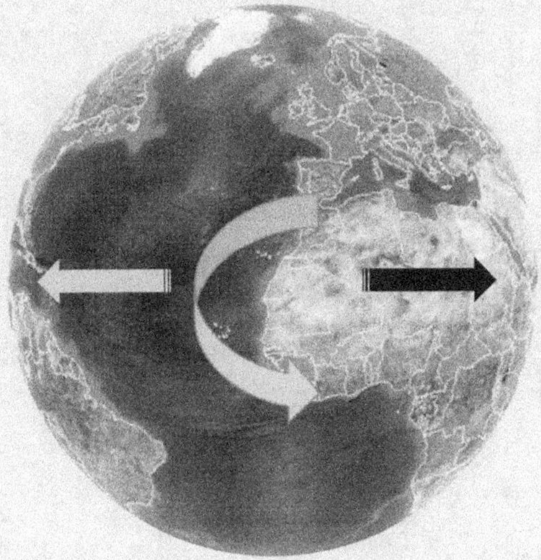

FIG. 1

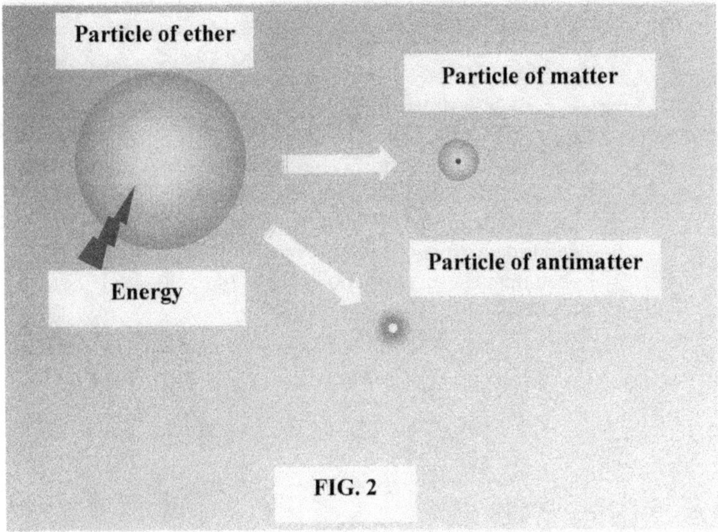
Particle of ether
Particle of matter
Particle of antimatter
Energy

FIG. 2

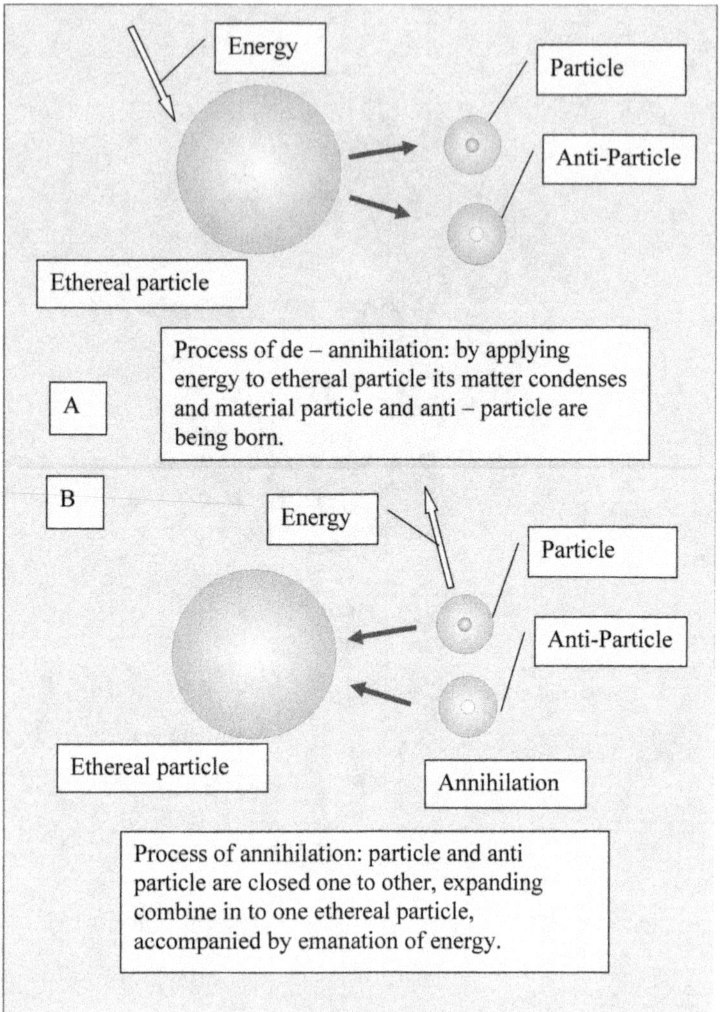
Energy
Particle
Anti-Particle
Ethereal particle
A
Process of de – annihilation: by applying energy to ethereal particle its matter condenses and material particle and anti – particle are being born.
B
Energy
Particle
Anti-Particle
Ethereal particle
Annihilation
Process of annihilation: particle and anti particle are closed one to other, expanding combine in to one ethereal particle, accompanied by emanation of energy.

Fig 3

ABOUT THE PHYSICAL FIELDS

There are a number of physical effects which still have no rational explanation, foremost – the physical fields, including gravity and the electromagnetic fields. The attempt to unite the four basic types of interactions – strong, electromagnetic, and weak and gravity is still without a result.

When an electromagnetic field acts in environment of deep vacuum, it is able to realize mechanical forces in the same way as without a vacuum. In both the cases these forces need supporting (applying) points and corresponding reactions, according to the third law of Newton.

The comprehension that there is nothing in a local volume, or in to the space of deep vacuum does not contribute for that problem to be solved. More, the gravity forces are active in a vacuum as well. It comes in to view other unsolved problem considering the earth or any other space body attracts others.

IS PHOTON A PARTICLE

According the contemporary comprehension, the electromagnetic waves within the part of spectrum that belongs to the light, are being translated by non-material particles, called "photons". Photon, according the contemporary theory has energy and impulse of motion but no mass. To prove the existence of photons, the equities of the energy and impulse are used (not shown here). From these equities follows that photon, being a material particle without a mass, may exist only when his velocity is equal to the speed of light and this is absolutely right. But, what from here follows is that according that statement the energy and impulse of the particle without a mass is undetermined (mathematically zero divided into zero). So, the proof of existence of photons hangs on an undetermined quantity. Within the further explanation we will return to this problem again from the point of view to the so called “wave corpuscular dualism”.

NEUTRAL AND ELECTROMAGNETIC FIELDS

The problem related to the physical fields (gravitation, magnetic, electrostatic and electromagnetic) is their unknown essence – namely, are they of some mass substance or not and how they work indeed.

The physical fields are explained theoretically at high average but it is sensible there are probable blank spots in relation of their rational explanation.

The official concept regarding the gravity still considers the objects in the space attract one another besides it is not explained how indeed. This explanation as the notion evolution are accepted officially at declarative level without making some profound explanation and thus turned into something like cliché.

The logical chain of the hypothesis is based on the Media accepting its meaning for all processes and as secondary effect giving unique opportunity to make rational explanations of what seems inexplicable.

Thus the basic link of the chain, the Media or ether becomes of importance and the physical fields to explain.

The physical fields are produced ever by the particles of matter. To this we add the real probability to be produced by particles of antimatter. In that way no physical fields are available out of the material world at least because of the lack of energy and motion.

There are two basic types of physical fields – electrically neutral and produced by electrically charged particles. The hypothesis asserts neutral physical fields are unable directly to interact with electromagnetic fields.

Up to this point of explanation we claimed Media is needed the physical fields to exist and to act. From this point on we are in the way to explain why this is so.

The reason is or at least seems to be very simple – each physical field is produced by material particle and after its producing it becomes a property of the Media. Media translates or keeps it.

Within the context of the hypothesis a physical field is any activation of ethereal substance by the particles of matter or antimatter.

Within that context a neutral physical field is produced by the covers of the material particles of semi condensed ethereal substance being apparently electrically neutral. These covers are the only part of the matter able to interact directly with the Media producing in that way a neutral physical field which we call gravitational.

In difference to this the electrically charged particles like electron emit impulse of energy in the space producing electromagnetic field. After its emitting the impulse becomes property of the Media which translates it.

A neutral field is indirectly a function of the mass of a particle, and independent to its charge. This is because the covers of semi condensed substance are function of the mass of the particle they are integrated to.

A neutral physical field might be a property not only of the covers of semi condensed substance but and of the particles of the Media. The hypothesis considers this as the only case when a physical field is produced not by material particles but by the particles of ether themselves. Further we will review this phenomenon within the context of the birth of matter and Big Bang calling it "auto gravity".

By contact of particles by their covers of semi condensed substance, as shown at Fig 4, forces (F) beating them off. Such forces cause for example the acoustic waves, when particles, such as atoms and molecules oscillate, heating the neighbors. By all mechanical interactions the particles of the stuff contact by their covers of semi condensed substance.

A presumption of the Hypothesis is that a particle without a mass does not exist. For example, it has been expected that neutrino could be such a particle without mass, sake of its properties, but late explorations led to opposite conclusion, besides, its negligible mass was measured precisely.

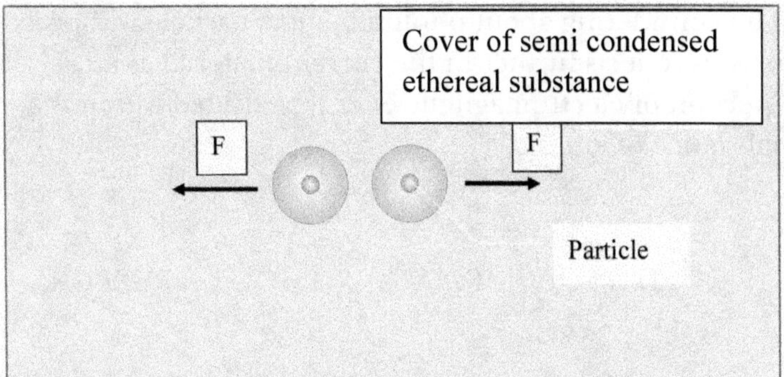
Cover of semi condensed ethereal substance
F
F
Particle

Fig 4

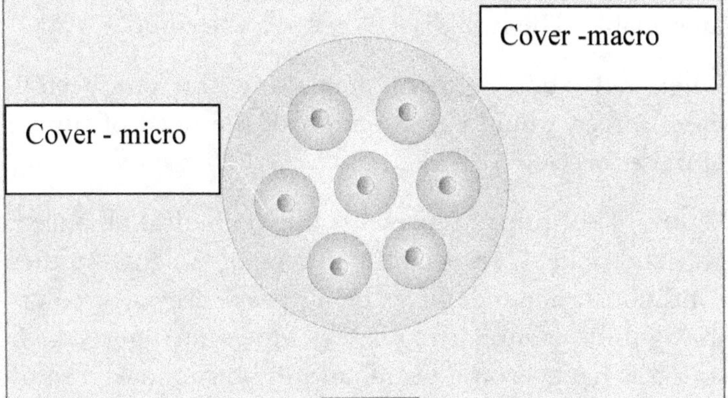
Cover -macro
Cover - micro

FIG 5

The considering about photon being a particle without a mass is reflects further in the current material as a "portion of electromagnetic energy, activated within the substance of ether".

COULD A FORCE TO ACT WITHOUT A SUPPORT

The motion on ice is a problem. The wheels of a car slip and it does not move, or if is in a motion, does not stop. The problem is a result of lack of support that normally is achieved by friction. According Newton's third law of motion an object experiences a force because it is interacting with some other object. In that case, the wheel or foot interacts with the ground by forces of friction, when these forces do not exist and interaction is unavailable. The result is – effective force could not act.

When talk about physical fields and the produced by them forces, would pay attention of the need of support of these forces.

Figure "The spring" represents a hypothetical absence of gravitation field, respectively, forces of weight. At these conditions a mannequin is being pressed by a force of a spring towards supporting base. At the same figure, A, the spring is hanged on a beam, and the beam locked to the same supporting base. At position B the beam is unblocked, there is no more supporting point of the link of forces, besides the spring is unable to press the mannequin. The system is mechanically free and each object of it should be freed to move in any direction.

Being activated, the fields create mechanical forces. As shown at Fig 7, wire (C), through which electric current flows, being positioned between the poles of a magnet is influenced by force (F). The force that an object exerts on another according Newton's third law must be of the same magnitude but in the opposite direction. That means the force that throws a wire aside must have a point of applying, reaction or supporting area. If we do not explain how the field creates that point, it would hang in to the nothing, simply, would not exist. The lack of it contradicts to the Newton's third law. But, the law is law, and must be observed.

It is the same, with the gravity forces. The considering that the earth, like any space object attracts others does not expound where the supporting point or area of these forces is. That expounding is a main topic of the hypothesis, and has a special chapter of it. At this point, a brief explanation of electromagnetic force follows.

The magnetic field between poles N / S of a magnet is created by corresponded charged particles, as resultant macro field integrity of their micro fields. The micro magnetic fields activate the substance of conducting media, the particles of ether, as shown at figure 8, A. Ether, being a substance of negligible inertia as conducting media, saves the magnetic field not influenced by the relative motion between the magnet and the media (Fig 8, A, B). Now, we are going to the electric wire, placed between both the poles. The electromagnetic field of the currency flowing through it interacts with the magnetic field between the poles.

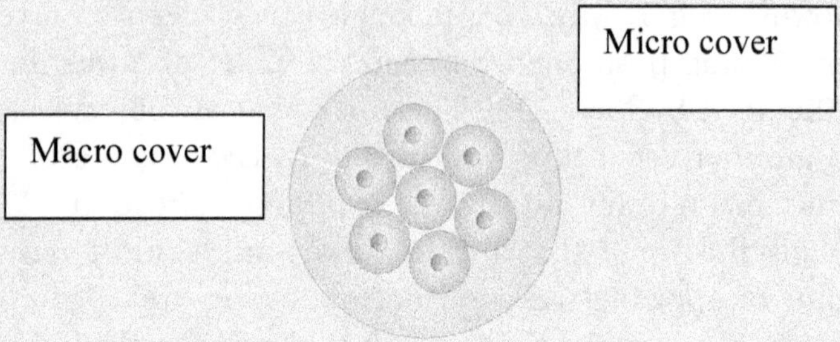

FIG 6

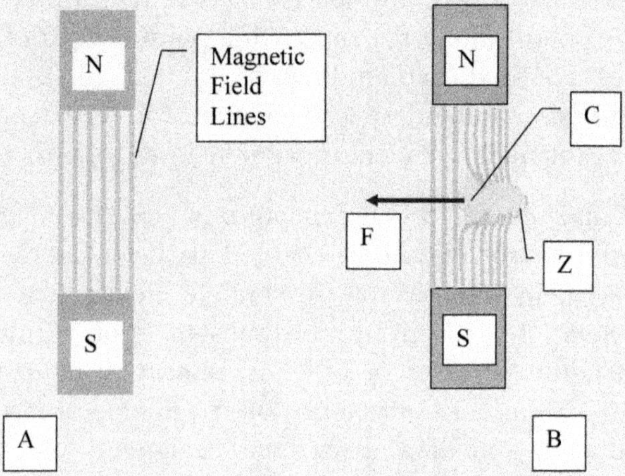

FIG 7

The hypothesis claims that each physical field is activation of the media or ether by the matter particles of semi condensed substance or electrically charged particles.

By the first case the field realized is neutral and corresponds to the gravitation. The second case corresponds to the example magnet – wire.

In that case we talk about electromagnetic field round the wire by flowing electric currency through it. According the hypothesis the essence of this field is activation of the media round the wire. The magnetic field of the magnet is activation of the media between its poles.

The activation is change of the state of the media or the ethereal particles which strictly corresponds to the parameters of the emitted magnetic and electromagnetic fields. These activations correspond exactly to the character and properties of the fields, their shape and intensity. Further considering of anything related to the physical fields of the magnet and the wire means the activation of the surrounding media. The interaction between bot fields is changing of the picture of activation of the surrounding media. Here we would seek and the supporting force.

The resultant force has its supporting local space, in zone (Z), where the ethereal substance is activated to higher potential and the lines are deformed, pressed one to other, action as a bundle. The lines could be compared to a bundle of bamboo sticks, tightly pressed one to other. But the physical character of the lines is activated media.

Let, for a while, return to the statement, ether does not exist and respectively there is no media.

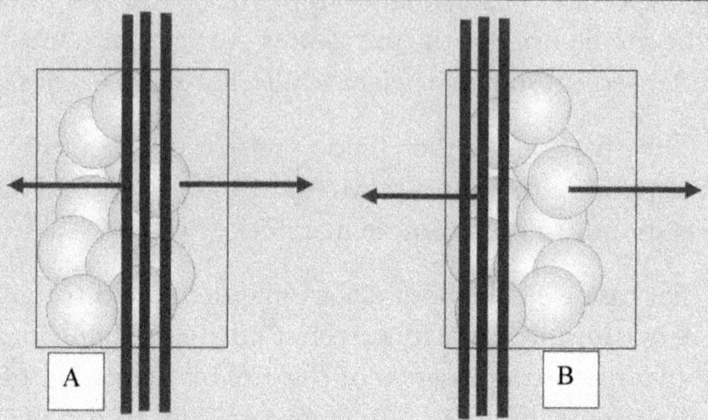
A
B

FIG 8

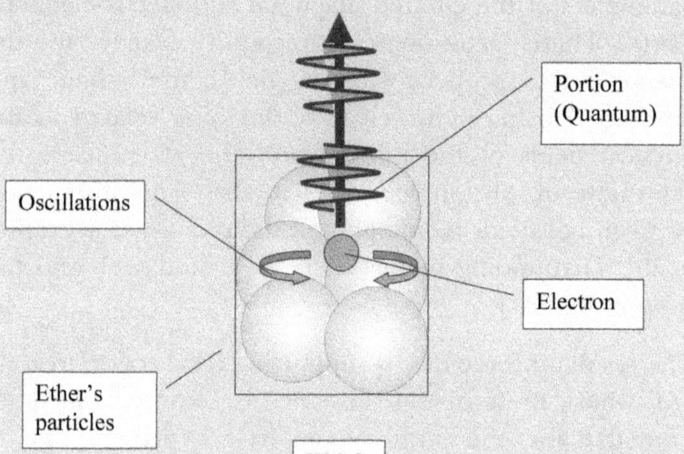
Portion
(Quantum)
Oscillations
Electron
Ether's
particles

FIG 9

Then, the magnetic and electromagnetic lines are not activated particles of the media but of some unknown other substance, because, by their creation, no matter from the magnet, respectively, from the wire has been separated, emitted in to the local space.

At figure 9 a charged particle, electron is activating the media. From the quantum theory, when it is accelerated by some source of energy, it jumps to outer orbit and emits a corresponding portion of energy, after which returns to inner orbit.

According the hypothesis, an electron, being accelerated to maximal velocity of oscillation, activates the neighbor particles of ethereal media, during the process of emitting. The media conducts that portion without inertia, exactly in the direction of emitting (vector). The velocity of conduction is limited to the known velocity of light, and the reason is that the electron is a material particle having mass and therefore inert.

If we have the chance to peep into other part of the material world, where hypothetically a charged particle with much lower mass than the electron exists, it could emit with a highest velocity. Hypothetically, this particle could be able to emit energy of hyper high parameters – speed and respectively frequency.

The creation and organization of the material world leads to the imagination, such an organization of the matter could really exist and we would suppose seeking it at the cores of the giant celestial objects. The presumption is, the matter inside to be at the edge between ethereal substance and matter. Such electrically charged particle of negligibly low mass hypothetically could emit impulses of hyper high frequency. We might suppose this particle at the core of a giant celestial object not to be engaged to some matter structure that is to say, in a free form of existence. Further, out of the core it might be engaged to such a structure as sub particle. But, being at a free form of existence it might be expected for this hypothetical particle to emit not discrete impulse but permanent signal.

It was expected light being a stream of non-material particles, but bold underlined – particles, according the contemporary concept, to exert mechanically, creating reactive force, or to act on a material body. But, the photons, having no mass could not to be obeyed to the second law of Newton. Photon's behavior does not meet the regulations of the classic physic, but only of the quantum. Without a mass, it has no cover of semi condensed ethereal substance. Photon is only bearer of electromagnetic energy of some properties.

The hypothesis asserts, any mechanical force is due to this semi condensed covers of the matter particles. The covers achieve contact between the particles and mechanical interaction.

Therefore, light's particles or photons could not exert mechanical forces. It comes directly the question to ask how the energy of the sun (as any celestial object) heats the matter of the planets. Apparently this is a transfer of energy from the star to the planet activating the particles of the cooled planet's matter.

Within that context light emitted by intensive motion of particles and electromagnetic impulses as consequence is turned back to motion of particles. But this is not due to interact between the particle photon with matter particles such as atoms and molecules. Particle photon if exists at all is unable to exert mechanical impact. The energy transfer is due to the reverse process – absorption of the energy by electron and putting in oscillating motion the particle to which the electron belongs.

As far as the hypothesis denies photon to be a particle but electromagnetic impulse the picture within its context is a little bit different. The impulse emitted becomes property of the Media. This might explain as an imprint within the ethereal particle which is temporary at the trajectory of translation of the impulse.

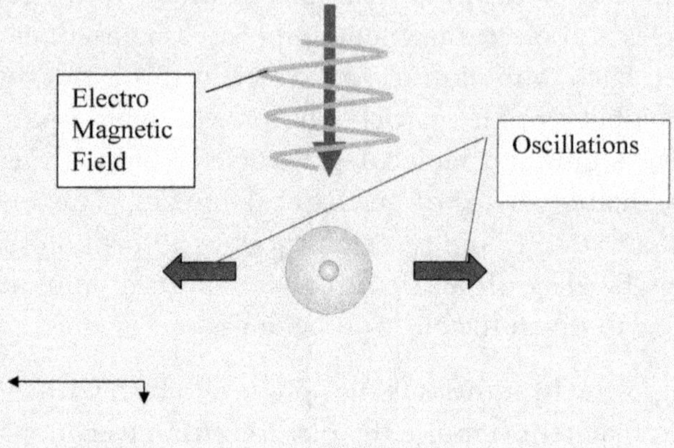
Electro
Magnetic
Field
Oscillations

FIG 10

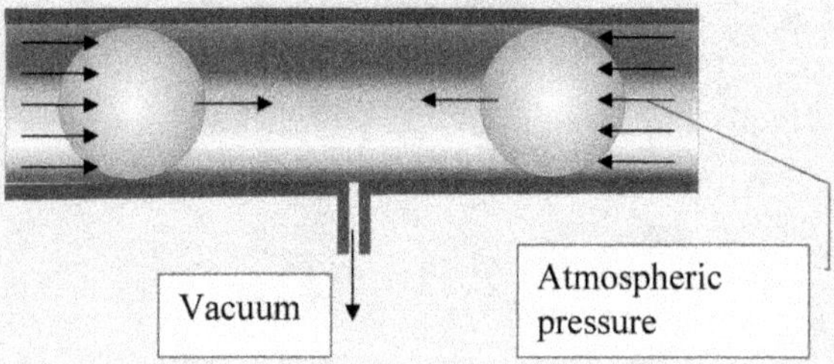
Vacuum
Atmospheric
pressure

Fig 11

GRAVITATION – RATIONAL EXPLANATION

According the contemporary concept, the space bodies are attracting one to another, and the forces of gravity are proportional to their masses, and vice versa to the distances between them.

The Hypothesis accepts entirely, absolutely all the mathematical apparatus that describes the gravity forces. The hypothesis concerns only the physical explanation. Before the explanation, a basic regulation of the hypothesis has to be formulated.

Forces of suction, pulling, and attraction of one body by another do not exist. The mechanical forces available are only of pushing / repulsing.

There are a lot of examples, when we say that a body has been "sucked" or "pulled". At figure 11 – A two balls are fitted in a pipe, and the air from the pipe evacuated via a hole. Both the balls are moving one to another; they are not sucked but pushed by the forces of the atmospheric pressure. At the same figure, position B, we would say, we pull the cart, but really, the force that has been applied is pushing the ring. More examples – the magnet does not pull the anchor, but the magnetic forces push it. The nuclei and the electrons are wrapped in to an atom or molecule by their covers of semi condensed ethereal substance. The proton, neutron and other particles that build the nuclei (the strong interaction) also might be considered as being wrapped by outer forces, not attracted one to other.

Weather a body attracts other is a subject that needs a proof, not only to be accepted as being evidence.

As shown at figure 12 body A, according the contemporary concept attracts body B with force "G", the weight. According the third law of Newton, the force of weight, as any other one needs support, or point of applying.

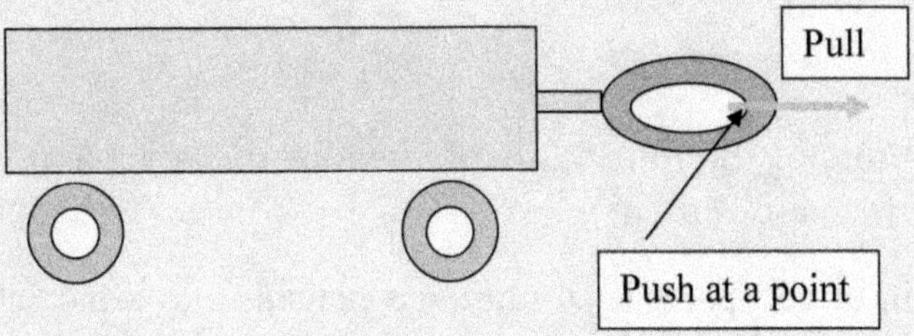

Fig 11 B

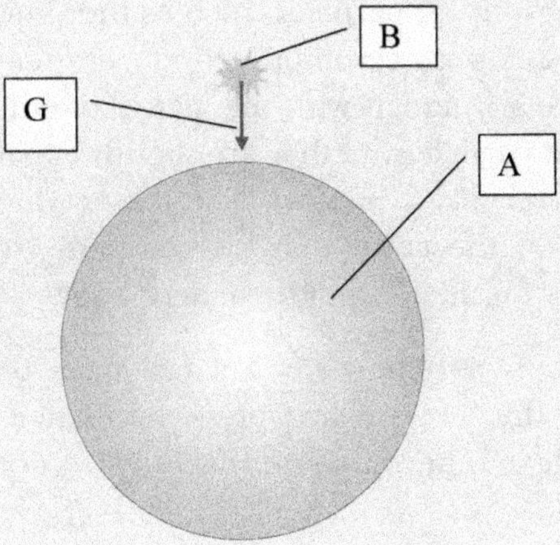

Fig 12

INTERACTION ETHERAL SUBSTANCE – MATTER

One of the four basic interactions is the gravitational, the weakest of them.

At figure 13 particle of matter is placed in to a particle of ethereal substance. Both the particles interact by their covers of semi condensed substance, function of their masses. Because of its negligible density the particle of ethereal substance is able to penetrate the particle of condensed matter, but there is a very slight interaction. As result of it, the ethereal particle is slightly expanded, the particle of stuff – slightly, negligibly pressed.

At that example, the interaction is at a level of particles. When a group of particles, such as the shown at figures 5 and 6 realizes common cover of semi condensed substance it interacts with the matter of ether in the same way – the particles of ether are slightly expanded, and the covers – reactively pressed. The forces which act depend directly on the cover of semi condensed substance and indirectly on the mass of the macro body.

Because of extremely low relative mass of the ethereal particles the forces which press the matter of a body are negligible and become essential only by very large mass of that body.

To proceed through the explanation, the space bodies would be considered as being big, large and giant space objects. Here a space object is considered as being small if it is comet, asteroid; big - planets, and large and giant – celestial object.

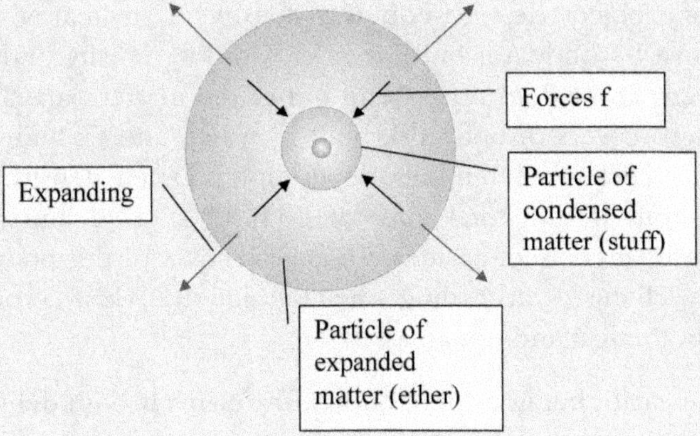
Forces f
Particle of condensed matter (stuff)
Expanding
Particle of expanded matter (ether)

Fig 13

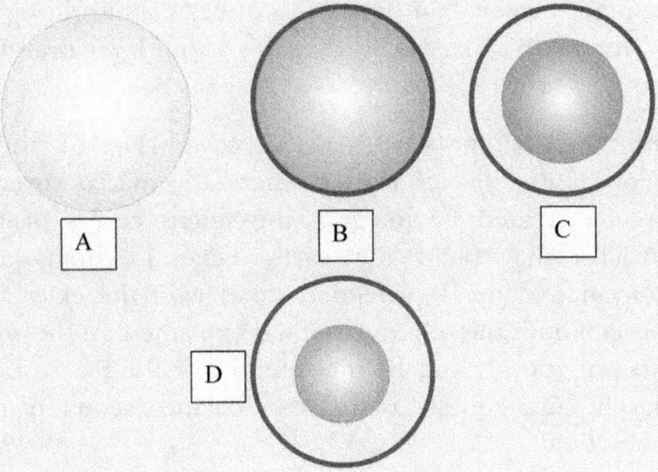
A
B
C
D

Fig 14

The common cover of semi condensed substance of any space object depends on the density of its matter. At figure 14, body A is built by gases with low density, and its macro cover does not extend out of the matter part. The macro covers of bodies B, built by water, and C, built by soil / rock are extending, depending on their density. At position D, the total mass of body C is saved, but the density increased, besides, the macro cover of this body is not changed according the regulation. This is only hypothetical and not an assertion.

The earth, being a space body with relatively high density of the matter that builds it has a macro cover extending out of it. If consider the earth as ideal sphere that it is not really, its macro cover is a concentric sphere with a diameter bigger than the earth's one. As shown at figure 15, the earth's macro cover creates a thin layer round it, about 3% of its radius.

The zones of the earth's macro cover (Fig.16) are as follows. Zone I – of matter, where the macro cover is directly created by integrity the micro covers of the particles that build the earth's mass – atoms and molecules. Zone II (extended zone) is of the extension between both the spheres, as it was explained. At the edge between zone II and III the intensity of the macro field sharply falls. Point C is the mass balance center of the macro field.

The macro covers have a crucial role by expounding the gravity physics essentials. In this context, the role of extended zone is very important for the live on earth. The gases that surround the earth have extremely low density, and from this – slight macro covers. The extended part of the macro cover being out of the solid stuff keeps the atmosphere of our planet from escaping out of it and contents the live, making it possible.

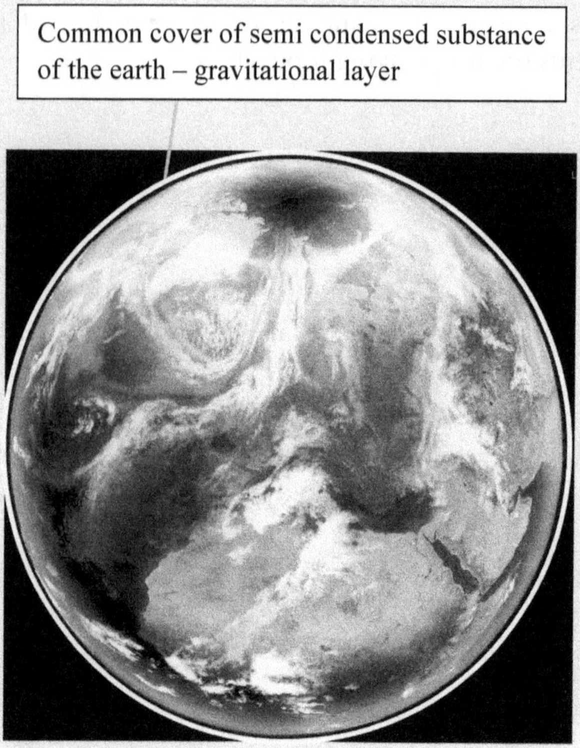
Common cover of semi condensed substance of the earth – gravitational layer

Fig 15

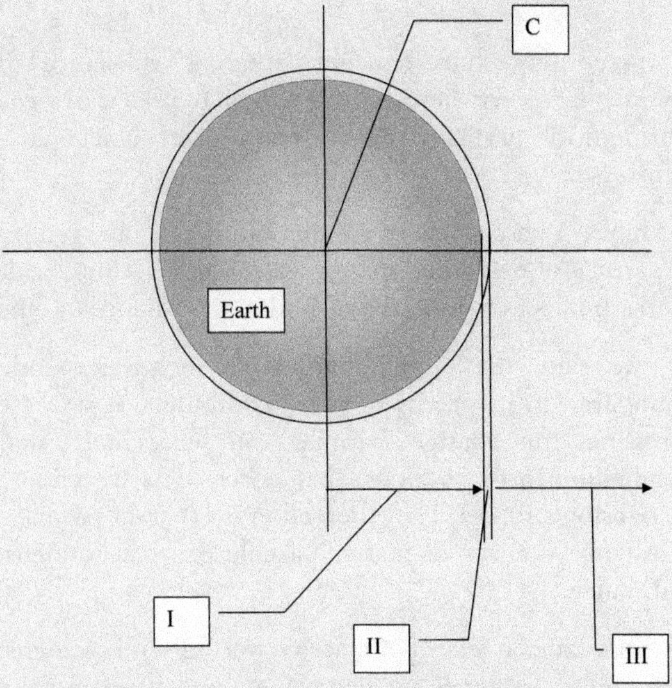

Fig 16

MATTER IMMERSED IN TO ETHER

According our concept ethereal substance fills everything everywhere in the space. It penetrates easily through the particles of matter and macro objects in the Universe.

At figure 13 a particle of condensed matter plunged in to a particle of expanded matter was sown. The first particle corresponds to matter, the second – to substance of ether.

It was said, the ether's particle is slightly expanded, compared to its size by nominal conditions, besides, it is pressing the matter particle with extremely slight, negligible force, besides that force is a reaction of expansion's forces. The interaction is ethereal particle or substance – cover of matter particle of semi condensed substance.

It is the same at a level of macro covers of semi condensed substance. An idealized macro body and its macro field, submerged in to ether are shown at figure 17. Couples of forces / reactions there act which slightly expand the substance of ether, and contrariwise, press the body of matter. By the presumption of ideal spherical macro body, the forces are targeted in to one point of balance (C) that coincides with its geometrical center.

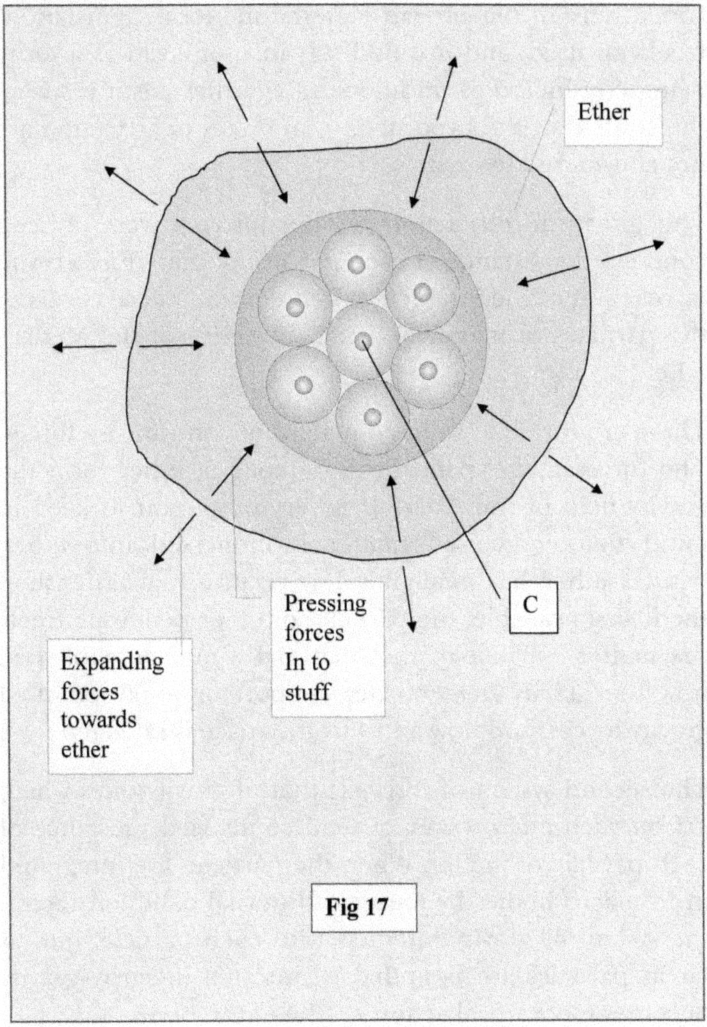

Fig 17

Before to proceed, we would make a stipulation and formulate a regulation within the context of the hypothesis.

Everywhere in the current material the term “gravitation” has been used, and it would stay in use, saved as a term, being considered as traditional. Regardless, as it is within the context of the hypothesis, real forces of attraction are not considered to exist.

The gravity forces action on the micro covers of semi condensed substance of each particle of matter as a result or reaction of the interaction between the micro covers of the particles of matter and the substance of the Media - ether.

There are formed two basic systems of contrariwise forces. The forces that expand the substance of ether form the gravity field of the space. It is very important to keep in mind that condensed matter is immersed into ether, besides ether has negligibly low relative mass (density), the lowest available, the particles of ether penetrate freely the matter. Further in that material would be explained how the gravity field creates support or apply point of gravity forces, and how and why the space is inflected.

The second system of forces is that of gravity ones which act on each micro cover of semi condensed substance of each particle of matter. When the particles are integrated in to macro bodies by means of internal cohesion forces, the system of gravity interacts with each particle, but, as far as particles are bounded by internal integrity gravity becomes mechanical action on the entire body.

The physics considers four basic groups of forces. These forces, also called interactions, are gravitation, electromagnetism, the strong force (a short-range force that holds atomic nuclei together), and the weak force (the force responsible for certain radioactive processes such as beta decay).

The intensity of interactions has been accepted to be expressed by means of a constant, a parameter without measure that reveals the possibility of processes, determined by the one of these interactions. A comparison of the constants, shows a relation between them correspondingly as 10 for the strong force; 10-2 for the electromagnetic interaction; 10-14 for the weak force and 10-39 for the gravitation. The conclusion is – the gravitation is the weakest of the interactions. Only the gravity fields of the large space bodies create essential gravity forces.

The reason is the negligibly low relative mass (density) of the ether – the substance that being deformed by the space bodies actions reactively on them, creating gravity field. Once again – the gravity field does not action on the bodies, it actions on the particles of the bodies. Only the integrity of the particles in to stuff that builds bodies leads to common action on them.

CREATION OF GRAVITATION FIELD

The subtitle of this chapter is "deformation of the space". When talk about mechanical forces, we are obeyed to take a look at the third law of Newton. Each force has equal in value and opposite in direction one. As in the case of electromagnetic and by gravity interaction we are obeyed to find application and supporting points and spaces of action of the fields and their forces.

At figures 18 and 19, the particles of ether are shown in an artless view, but that simplifying is needed for the purpose of explanation. There are distinguished the next zones as follows.

Zone IZ (Internal Zone) comprises the earth's macro cover of semi condensed substance (the same explanation is valid for anybody). In this area the space object's action on the ether's substance into which it is being immersed in a given moment is as a giant magnifier expanding the ether's particles. That expansion is most in the center point C, and reduces by increasing the distance out of it. Generally, in zone of the macro field of a body particles ethereal are being expanded over their nominal measures. It may be said, the total substance of ether is expanded, deflected.

Zone OZ (Outer Zone) starts from the edge of macro field of a body and stretched far from it. In that area, the particles of ether aim to compensate their deformation from the internal zone, and to approach a status of internal balance. Smoothly, they pass through their nominal measure and continue to decrease it. Zone OZ ends where the particles reach again to their nominal measures.

The basic deformation is in the space of a body's macro cover of semi condensed substance, and the deformation in its outer zone is an effect of it. As shown at figure 19, the macro cover of the earth (valid for any big, large, and giant space object) acts as a giant lens. The earth, as any space object is in a relative motion towards ethereal substance. Because of this, the process of deformation and "lens effect" are not static, but strongly dynamic. That could be illustrated by means of a lens being put in a motion over particles that correspond to the motion of the macro cover through the media.

The deformation of ethereal substance is a result, as was mentioned above, of interaction between micro fields of both ether and space body's particles. That deformation leads to misbalance and internal tense that determines creation of gravitation field. At this point we can state, space objects do not have nothing as "own gravity field". Their matter causes its creation, being immersed in to ethereal media.

This explanation is based on the supposed interaction between matter and ether. By this explanation each matter particle expands slightly the substance. This to be compensated might be expected somewhere within this system of interactions the substance to be condensed. Really we might consider the particles of ether to be closed to their nominal state at the mass center of the space object and slightly pressed outward following linear function. In that way of reasoning the particles of the media from the mass center outward become deviance from their nominal state which might imagine as ellipsoids which small axes pointed to the mass center.

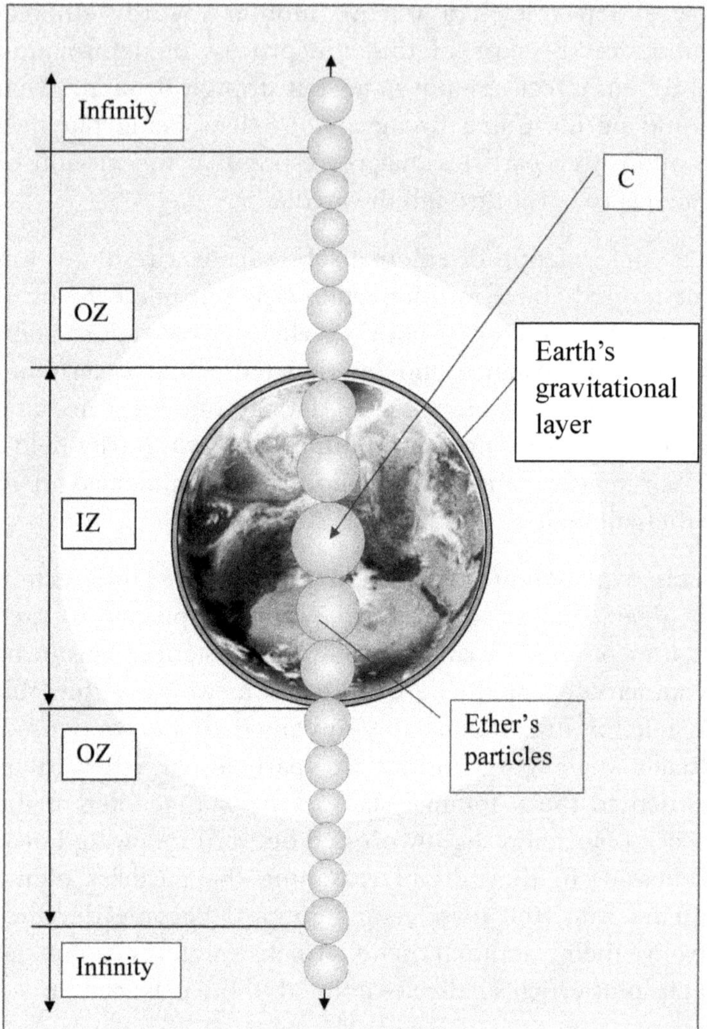
Infinity
C
OZ
Earth's gravitational layer
IZ
Ether's particles
OZ
Infinity

Fig 18

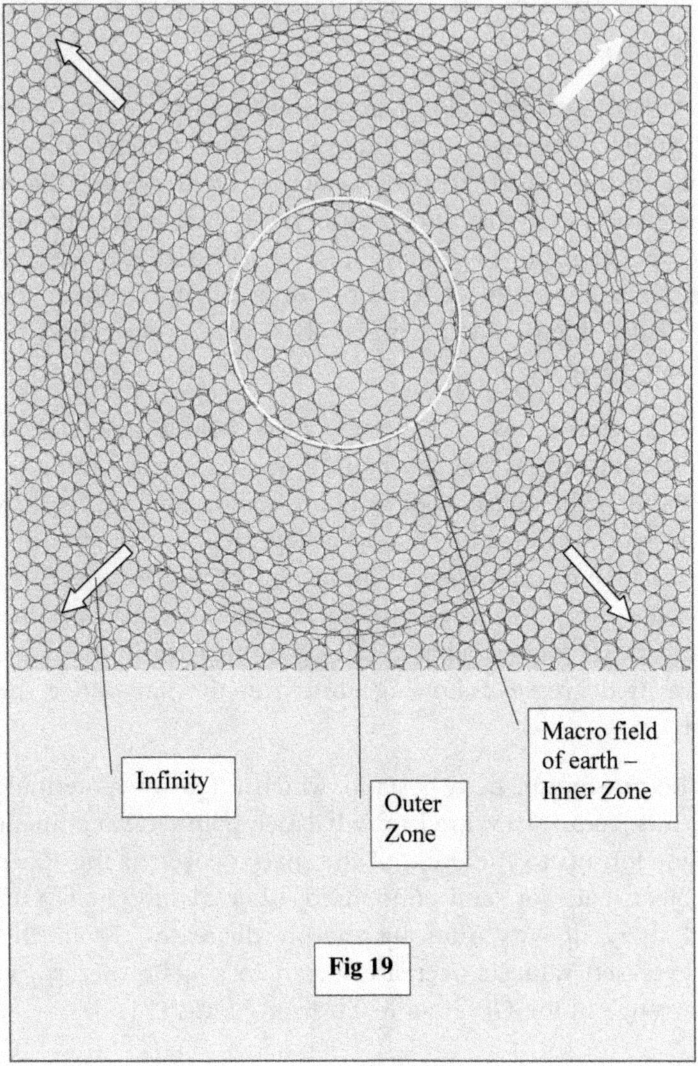

Fig 19

ZONES OF A GRAVITATIONAL FIELD

The gravity field acts on each particle of given material body, and the integrity of the particles in to a macro body leads to general action on the entire body. The forces of that interaction have their applying and supporting points determined by pressed particles of ether being in the outer zone. Because of extremely low density of ethereal particles, that zone is stretched far from the space body.

There are zones of the gravitation field that could be distinguished as follows: zone that comprises the matter of which the space body is being build; zone defined by the space extension of macro cover of semi condensed substance of the object, if exists. Out of the zone defined by the cover extension the activity of ethereal substance sharply decreases. This defines the third zone where the gravity decreases following linear function far out of the object.

The gravitation field is static, which intensity (potential) grows from zero value in the balance point "C" at a linear function up to the edge of the macro cover of the space object made of semi condensed ethereal substance. Out of that edge its intensity sharply decreases. From this decreased value it decreases again to a value of zero, at the edge of the Outer zone, far from point "C".

When the space object is of a spherical shape, the equipotent surfaces round of it are concentric potential spheres. Indeed, they correspond to the stage of expanding / condensing the particles of the ether's substance, as was described above.

Largest space objects, the celestial ones are relatively independent, not being in orbital motion round others. They have giant sizes; their matter could be highly condensed, of high density. That determines extremely large macro cover of semi condensed substance and deformations of the ethereal media. Respectively, the gravitation fields round such bodies are spread far in the space.

As shown at figure 20 A, both the stars, being closed at a critical distance, are pressed from outside by forces FO. These forces are created and supported as was described above by deformed ether's substance, respectively, a common gravitation field. The ethereal substance between points C1 and C2 is being pressed (at figure – gray zone) and reactive forces interact and to keep the bodies at a distance.

A small object is dependent to a large one – if it is in orbital motion round it. In this case, the whole gravity field of the small one is immersed in to the field of the large one. This explanation is in principle; the system of forces that interact is really more complex.

As shown at figure 20 B, the gravity field of moon is immersed in to the field of the earth. The ether's particles are pressed once by the mass of the earth, and again – by the mass of the moon. Both the bodies are pressed by outer forces acting towards the direction that connects their mass centers, caused by the said reason. The ether's particles between earth and moon are less pressed. The aim the moon to fall on the earth is compensated by the centrifugal force of orbiting the satellite round the planet. The zone of relaxed substance between both the objects causes the influence of the moon on the earth's gravity field.

At figure 21, using the lens effect, the space deformations are represented, the schemes have illustrative character only.

It is worth to explain, why the space object's orbits are so precise. The reason is the interaction between the micro covers of semi condensed ethereal substance and the ones of the media. The integrity of particles of the bodies fosters an illusion of interaction between them; indeed, it is at atomic, respectively, at molecular level.

How a large body (consider the earth) "attracts" a small body. Being in a zone where the macro field of the earth extends the sphere of the earth's stuff, a small body is immersed in a strong gravity field, caused by the deformation and intension of ethereal substance. Each particle of that small body is pressed by forces of that field, exactly and precisely towards the mass center of the earth. A body, being free in the space is being put in a motion by these forces.

Because that force is namely body's weight, there is a concept, when a body free falls to be in a status of "weightless". But the forces act, and nothing could cancel their action, what means, the "weightless" is not correct. The "weightless" is simply a physiological phenomenon. More – we say "up" and "down". This is also an effect of a physiological sense and orientation. In the Universe there is no “up” and no “down”.

A practical example, from location of technology, related to that topic is the forming of rolling bearing’s spheres. These spheres, being perfectly formed beforehand, by high temperature of metal, are being cooled to normal temperature falling from high towers in a vacuum environment. By falling, they save their ideal shape, because, the gravity forces action on each of their molecules absolutely equally that guaranties their shape perfectly to be saved.

The hypothesis asserts the particles of the media interact with this part of matter which corresponds to their essence. This part is the cover of semi condensed substance of each material particle, atom, molecule, and of each macro body. The covers define the interaction.

For a body in the gravity field, the particles of the media are more strained at its outer parts than to its inner parts towards the mass center of the earth. This is a potential difference defining the gravity force which aims to move the body towards the earth’s mass center.

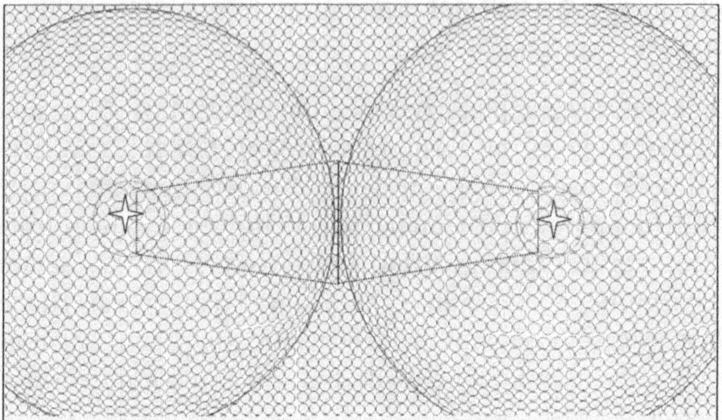

Fig 20

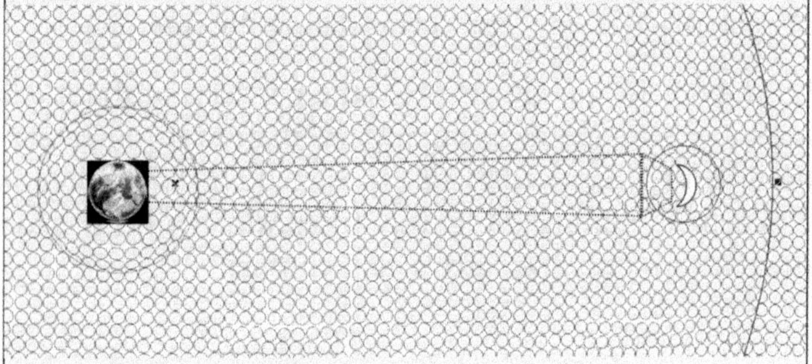

Fig 21

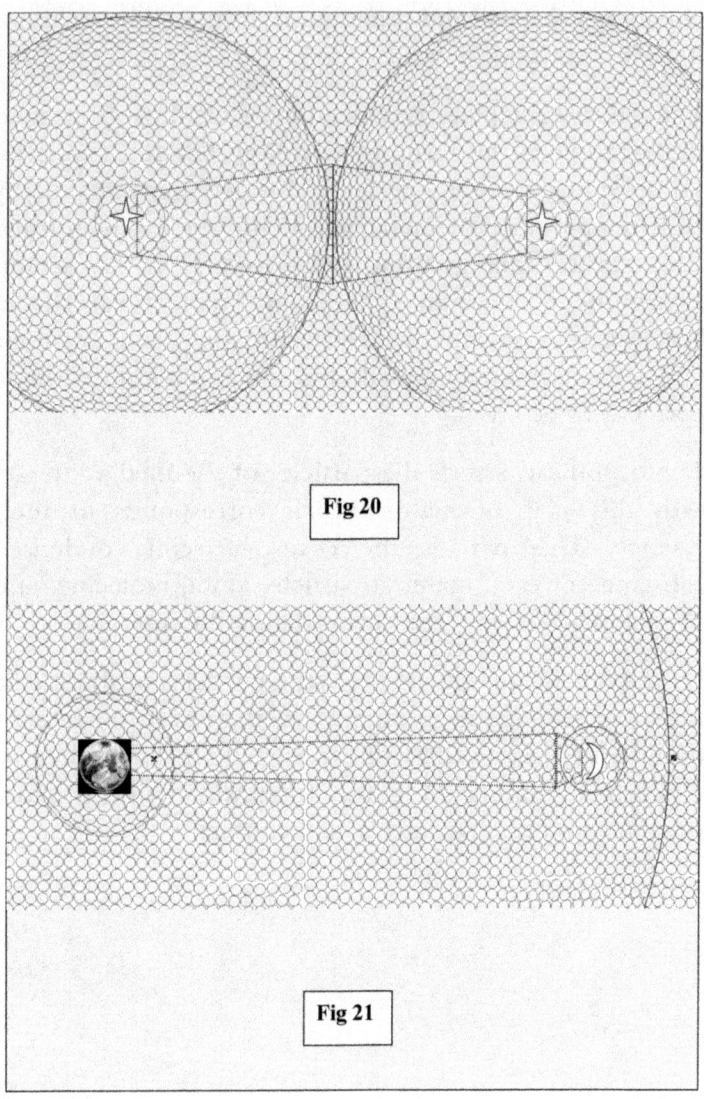

EFFECTS FROM MEDIA DISTORTION

If we say, ether is the space, namely, to identify the space with the ethereal substance that fills everything everywhere, it would be correct to consider space as being deformed, because of influence of large celestial objects immersed in to it. That deformation leads to effects that determine not only the gravity fields creation, described above, but and to alteration of its conducting properties. Regardless the gravitation forces are weakest among the interactions, large celestial objects that create a system in the Universe cause essential distortions of the ethereal substance.

One of the questions is: is ether motionless, or in a motion, is it available to put somewhere an absolute coordinate system, from which to account all the processes. Currently, seems, there is no answer, because the ether's substance is not detectable by the material facilities, and could not be explored.

The second question is about, how that deformation affects the conducting properties of the media. The current concept is light is being translated through an empty space by material particles with energy and impulse of a motion, but without a mass. The photons lead to problems with the wave character of light, including interference and diffraction to be explained. It is well known, light is visible in to the space only from the view point perpendicularly towards the direction of its translation (vector). From any other position, it is invisible, the space is dark. Could be expected, material particles to be visible from any view point, and the space to be illuminated.

The hypothesis explains the light as discrete portions of energy, translated by the media of ethereal substance, activated by electromagnetic field of charged particles (electrons). The concept for ethereal media puts the things in to a logical order and gives a chance to solve the said problems. Additionally, the hypothesis does not accept material particles without parameter “mass”.

The contemporary concept regarding the influence of light by the gravity field is as being a result of the same forces that action on each material particle. Being material particles, photons “fell down” in environment of a gravitation field. Being considered as material ones, the photons must have the same behavior. But, the gravity field interacts only with particles with covers of semi condensed substance, indirectly, mass, a property that the photons do not have. The gravity field indeed influences the light, but it actions as a distorted conducting media. That influence is slight, because of the extremely low relatively mass, high penetrating and conducting properties of the media. Its action is essential at very long distances in to the space. Here is the so called “gravity lens” by which phenomenon on light act not the gravity of large space objects but the distortion caused by it. In addition, any distortion of the space could change the light’s trajectory.

INERTIA OF THE MATTER

There is one more a question, regarding the forces and masses of the bodies. This is the difference between inertial and gravitational mass. Difference does not exist; this is exactly considered by the science. But, there is a principle difference between force applied to a body and the acceleration, caused by it. Let distinguish gravitational and mechanical forces. The gravitation force actions on each particle's cover, each other mechanical force actions at a point, or a spot of a body. According the second law of Newton, acceleration depends on the force which acts on a body and on its mass. Here appears a distinction between both the actions – the acceleration caused by gravitation forces does not depend on the mass of a material object.

Being caused by a neutral physical field, gravitation force actions on each mass particle of a body, namely, on their neutral covers. Further, respectively to the integration of body's particles in to a hard material, the gravitation micro (elementary) forces are as well integrated and create a common force that actions on the whole body.

In that way, the integrated force depends on the mass, and automatically compensates any gradient of mass. It is in its essence an extensive quantity, additional mass automatically adds amount of gravitation force. The acceleration depends only on position of a body in the potential sphere of gravitation field.

The acceleration when a non-gravitation force acts is a function of its mass, because the force actions on the whole body (its mass).

The basic form of matter is the substance of ether that fills everything everywhere in the space. By a process of de annihilation, the antipodes that build the ethereal particles are being separated. Leaving their previous form of symmetry and balance, the annihilated particles aim to create new structures, to approach again a balanced status.

The hypothesis considers, the process of mass producing could be initiated in some point of the space by conditions of a large deformation and powerful intense between ethereal particles as consequence, or by the influence of an energy source. The parameters of both the influences must be at a level to be able to origin, keep and accomplish a chain react. When that requirement is not fulfilled, the separation of mass producing particles is instable, and their live – short. In that case, they return to their previous status of balance. When the local conditions make possible a chain react to be realized, it accomplishes with an explosion, besides a body of plasma is produced. This is the first stage of creation of new structure, where the particles aim to approach a balance form of their existence. No more they could return to the previous status, because the reaction, being supported by powerful local intension and energy, is irrevocable.

There are three consequences of the process in interest - the matter, the antimatter, and the space, where the reaction has been performed. Further, the matter builds a space object of plasma. But, the behavior of antimatter could be only a subject of imagination. Being separated from the matter, could build other a structure, where to approach its status of balance. The behavior of the antimatter, as well, the deformation of ethereal media in a local space, where the process has been performed could be in a relation to the dark holes in to the space. The structure might be organized as a radial symmetry in the center point of which the dark hole of antimatter is positioned. This is not assertion of the hypothesis but only a consequence of a considering.

FORMING OF CELESTIAL OBJECTS

The Hypothesis explains how a body is being influenced by the gravitation field, respectively, forces. Each elementary particle of it, floating among the particles of deformed substance of ethereal media experiences the intension of that deformation. As a result, elementary forces act on the particles in direction towards the center of balance of deformed ethereal media. The integrity of these elementary forces and their action create a gravitation field, which intension in each point corresponds to potential of the field in that point.

When a body is structured by elementary and complex particles in some integrity, the elementary forces of a gravitation field create a resultant force applied in to the mass center of it, action in the same direction as the elementary forces.

A basic influence of gravitation fields is that by shaping of celestial objects. The cloud of plasma, created after an explosive mass producing, as it was explained, has no determined shape. That cloud deforms the media and creates a gravitation field round of it. Being pressed by gravitation forces, the particles of the cloud are put in a slow motion towards its temporary mass center. As shown at figure 22, where a plasma body is represented, the forces that action along its wide side (FW) are with a superior strength than the ones, that action along the narrow side (FN), because the media is differently deformed by both the sides. The shape of a sphere, formed finally, is the only available for the forces and masses to be in balance. If the body is being cooled before the spherical shape to be obtained, gravitation forces are not able to form it more, and it saves a non-spherical shape.

At this example it is seen no body attracts other, but elastic forces of deformed ethereal media press it to obtain the shape.

By the motion of large space objects or by explosions and birth of new celestial object disturbances of the space occur. If the space in the Universe is empty or filled by occasionally or permanent streams of particles only, there is no media these deformations to translate. Introducing the primary substance ether by the presumption it fills the space everywhere in all directions we have that Media and become able to explain the gravitation waves and the space distortion in the Universe.

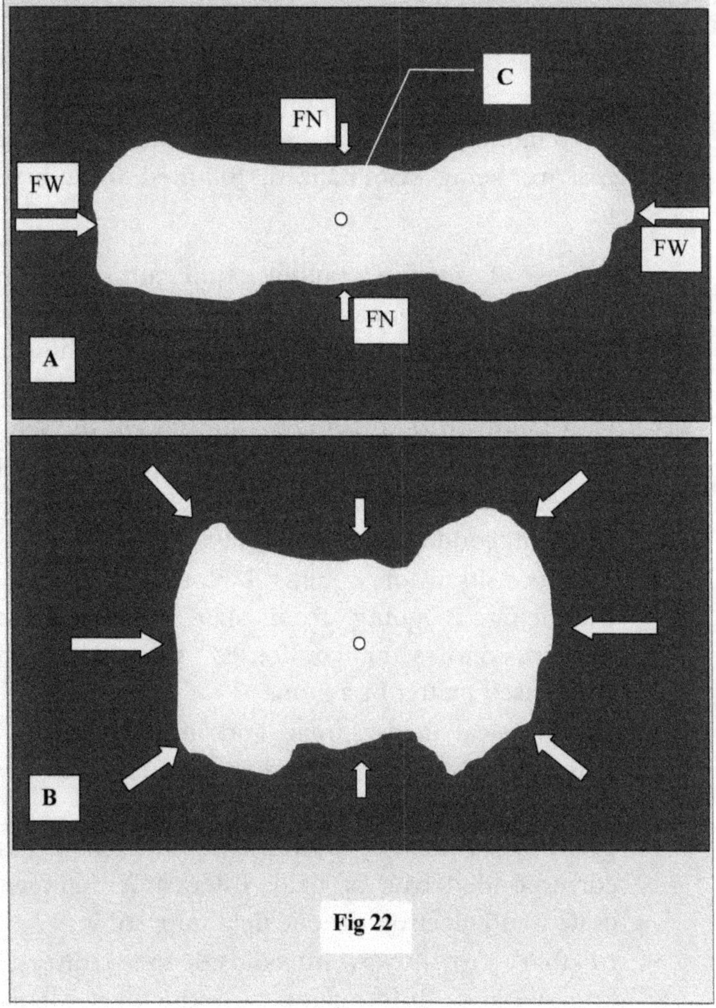

Fig 22

GRAVITATION SUMMARY

In this chapter, the regulations, introduced by the hypothesis are being systematized, followed by a brief comment.

- Forces of suction, pulling, and attraction by interaction of material bodies do not exist. Mechanical forces available are only of pushing / repulsing character.
- Gravitation forces act on the cover of semi condensed ethereal substance of each particle of matter as a result of interaction between ethereal and semi condensed ethereal substances
- The transition of a mass by process of mass producing is going from state of expanded substance to semi condensed substance and condensed matter by a jump.
- The physical fields spread and interact through ethereal substance becoming property of the media.
- The physical field of a given type interacts with a corresponded type of field. Interaction between gravity and electromagnetic field impossible.
- In that part it was introduced the Universal Energy and Information Conducting Media (UEICM). The features and functions of that substance are systematized as follows.

- It consists of a substance of negligibly slight relative mass (density) and thus penetrates the matter everywhere, fills everything.

- The particles of the substance of the Ethereal Media do not have energy and impulse of a motion being in a status of relative internal balance

- The particles of the substance of the Ethereal Media are being activated by the micro, respectively macro covers of semi condensed substance besides conduct the energy and information of the fields

- The substance of the Media, being activated by electromagnetic fields generated by matter creates supporting points / spaces of forces, caused by these fields. The Media has negligible inertia by all the processes, including penetration and energy conducting; does not influence the particles and structures of matter by their relatively motion; is able to conduct energy and information with a frequency and velocity much higher than the velocity of the light

- The energy and information conducted by the substance of ethereal Media with a frequency and velocity much higher than velocity of light is being emitted by unknown particle and undetectable for material facilities

- The substance of ethereal Media is being deformed by the space objects being immersed in to it; its deformation is accompanied by internal tension pressing material objects immersed into it

- The big large and giant space objects only cause essential deformation and intension of the Media creating round of them essential gravity field

What is the force that attracts us to the earth? Indeed, that force presses us as shown at the image. It needs support according the third law of Newton. At the figure, position (A), the spring presses the mannequin to the ground, being steadily attached to the basis by means of a console. At position (B) that attachment is being released and the man free to move in any direction.

At the next image the particles bellow and above the man are respectively expanded and condensed creating potential difference and a force that pushes the mannequin towards the basis. Indeed the particles and above and below are equally tensioned and action on ones of the material body pressing them to the center of the earth

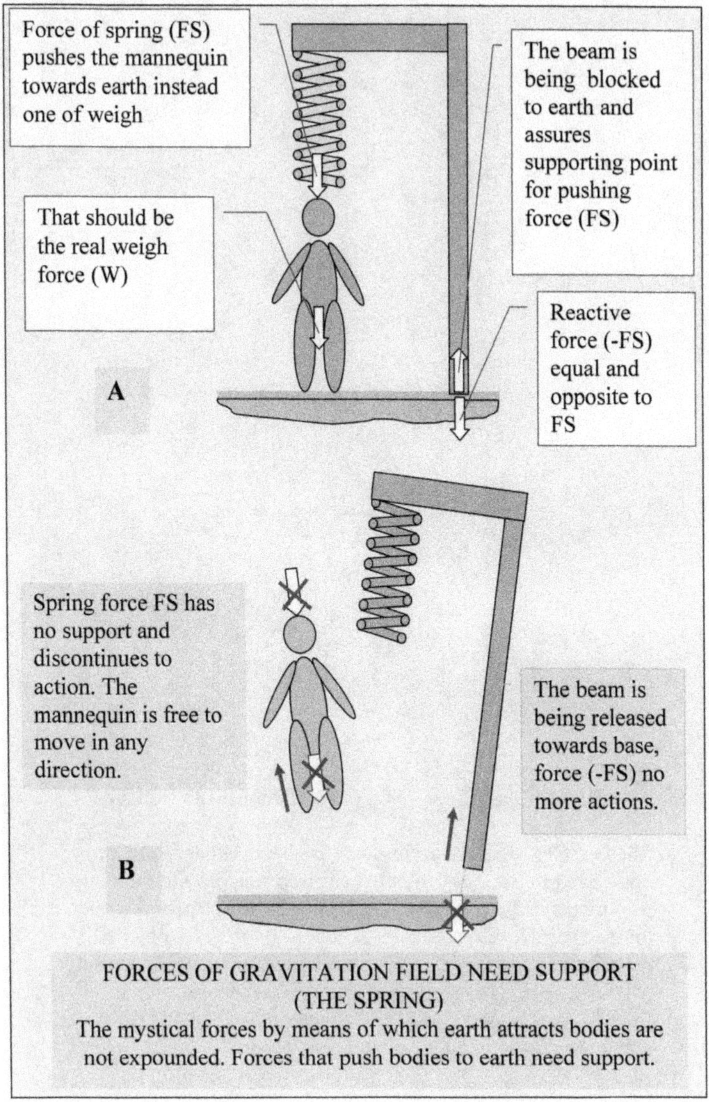

FORCES OF GRAVITATION FIELD NEED SUPPORT (THE SPRING)

The mystical forces by means of which earth attracts bodies are not expounded. Forces that push bodies to earth need support.

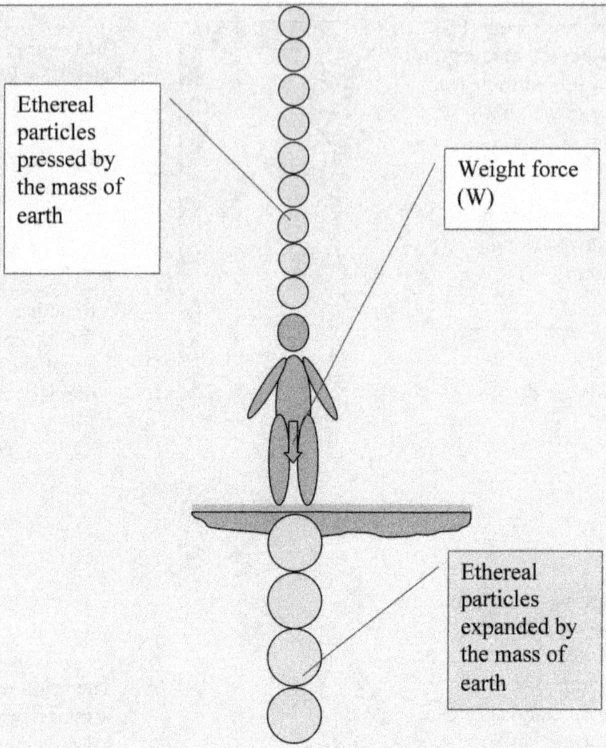

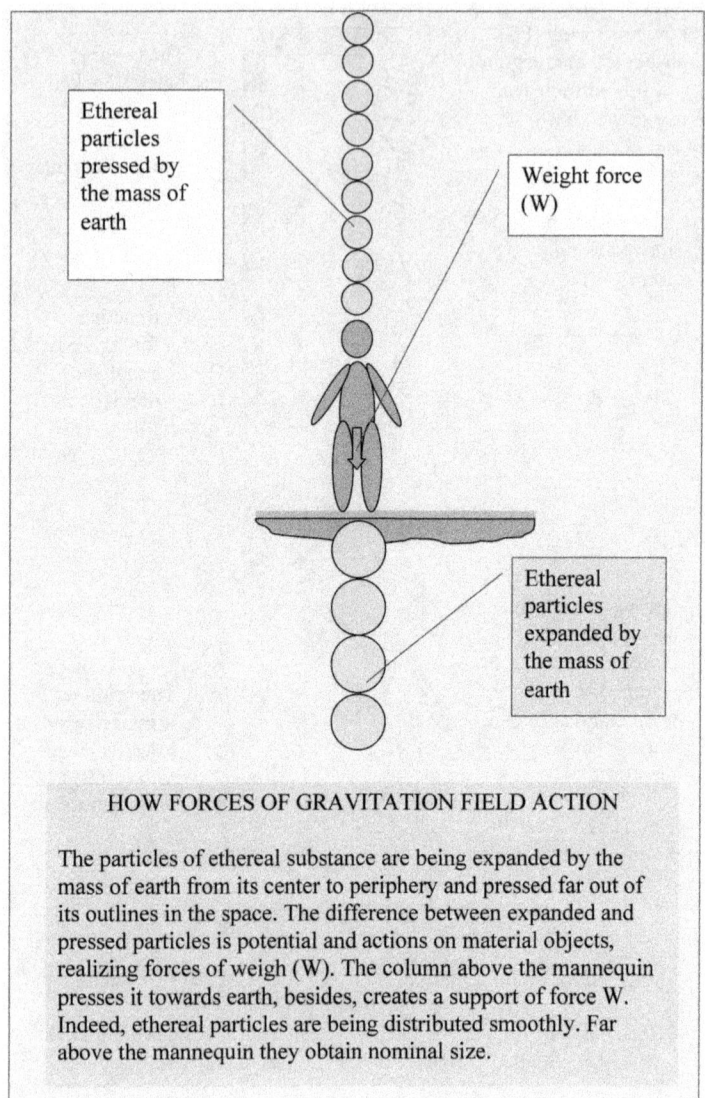

HOW FORCES OF GRAVITATION FIELD ACTION

The particles of ethereal substance are being expanded by the mass of earth from its center to periphery and pressed far out of its outlines in the space. The difference between expanded and pressed particles is potential and actions on material objects, realizing forces of weigh (W). The column above the mannequin presses it towards earth, besides, creates a support of force W. Indeed, ethereal particles are being distributed smoothly. Far above the mannequin they obtain nominal size.

LIFE

SCIENCE, SCEPTICISM, PARANORMALISM

What we are able to see, to touch, and to explore is what really exists, we are considering. Everyone from the middle Ages would be really surprised to the opportunity to watch TV, to talk by a mobile phone. No one would believe in the probability of the technology wonders of our civilization before to see and touch them. It should be a problem to understand that physical principles are being the reason these usual today devices to be reality.

The science has developed a perfect mathematical apparatus that describes the physical phenomena and processes, including these, related to the gravity field and forces. In that concrete case the result of that mathematical description corresponds to the Law of the Space Attraction between material bodies.

As it was postulated in the first part of that material, the bodies are unable to attract one to another and there is no evidence, proof, or rational explanation how really this is available. Here is the question, what is more probable, the attraction without explanation, or the expounding from the first part, based on a logical chain of reasoning.

Is it possible or permit to that hypothesis to contradict to unproved, but deposited from centuries concept. More, what about the concept for existing of ether, the universal conducting media for all alien communications, through which the principles of gravitation were expounded? Let do not forget, the existence of that media has been supposed first, and rejected next by the most great physicists.

Regardless of the rejection of the Universal Information and Energy Conducting Media, the ether, it presents and in the second part of this material, dedicated to life and evolution.

The mankind has acquired the understanding, when somebody is in possession of some unusual abilities for communication or others, this is paranormal.

In its nature, all the religions with their rituals are being directed to God, who has us created, and Who determines our destiny. But, all this is immersed in to foggy, mystical belief. Are miracles possible, or all this someday could be set into a frame of a rational physical explanation? Are we allowed to peep into that unknown and mystical today world, and if yes, to where? Are we able to expound who we are and Who has us created?

The hypothesis does not reject anything, does not give directions. Simply, it follows a train of reasoning and confesses, miracles are unavailable and do not exist, everything is based in a logical way to the laws and regulations of the physics. But the hypothesis not in the least does not reject the spirit, idea and the word. On the contrary, it finds a correlation and harmony between the material and spiritual.

When talking about normal and paranormal, the author remembers his collie dog Aster and how has been impressed by the dog's behavior and abilities which has watched for long years. The dog was ever perfectly informed about who is at the other side of the front home door. It was able to distinguish whether there was a member of the family, unknown person, or other a dog. It was impossible to sense the smell through the door. The dog knew when a member of our family is coming being at long distance.

There are known the abilities of the animals to predict change of the atmospheric and seismic conditions. In the same way, people having optical or acoustical problems are hypersensitive by their unaffected senses. They feel the presence of others without seeing them.

All these cases we accept as being normal, we do not consider them as being paranormal.

In that context the Hypothesis considers everything as being normal and obeyed to the physical laws, principles and regulations and gives an rational explanation how and why. In that context would be expounded how a men breaks tea spoon's grips not as being paranormal, but as being distinguished compared to all others.

ENERGY, SPIRIT, MATTER

The hypothesis is restricted within a specific frame of validity, not forgetting it is a hypothesis only. The area of validity is being determined by the principle of abstract and deductive reasoning. Beyond that frame we should only imagine what might be.

As it was stated in the first part, “Gravitation”, everywhere in the space the Universe is being filled by elusive substance. When we take a volume of air and empty it, in the formed space of vacuum would not to be nothing, or occasional particles of matter, but that volume remains filled by the substance of ether. The same is about the Universe, it is filled by the expanded substance of ether and we are being immersed in to it. The expanded substance of ether has a big volume of its particles, they are motionless, and respectively have no impulse and no energy. With these characteristics, ether is universal conducting media for all the physical fields, regardless of their parameters.

Let imagine, somewhere, there is a matter, so structured that to be able to radiate electromagnetic waves with frequency many more times higher compared to the ones of our material world. Also, there might be a transformer of energy able to transform the frequency of our energy to the hyper high value. These waves are being translated in to the space with a velocity many more times higher than the velocity of the light. The Universe is being pierced by these information and energy translated waves.

The religion has a main idea and it is: the spirit is a base of bases. The Hypothesis does not contradict to this, but claims, the spirit uses the physical media, laws, regulations and principles to be spread everywhere.

CREATION OF BIOLOGICAL STRUCTURES

The charged and neutral particles packaged into atoms and molecules build the matter that builds all around us in our material world. This is the inorganic matter; it is a part of the processes before the appearance of living organisms and after that. The development of the inorganic matter is result of physical processes depending on geographic position, phenomena and changes of the environment. Such are the cooling of plasma in the upper earth's layers, temperature changes, water, damp, electromagnetic and other physical fields and others. As a result matter becomes a structure closed to what we see today.

But, never have learned or heard the particles of the inorganic matter to be obeyed to laws and regulations distinguished to the physical, of course, the ones of the quantum mechanics. Never have learned or heard these particles to give evidence of some form of intellect in their behavior.

Having in mind the extreme complexity of even the simplest organism and the impossibility one to be synthesized by the contemporary conditions and technology development, should be relevant to ask, is it probable the particles to be self-organized in so complex structures somehow of itself? When distinguish the matter to organic and inorganic, let ask about something more simple, is it possible the particles to have some logic to create more complex organic structures.

Consider, the influence of lightning as being such a condition, an organic molecule to be created, that structure must survive, to be reproduced, and finally, to reach to a living organism, some complex logic has to integrate these molecules in to an order. Nothing like this could be expected, the organic molecule is instable, and on the elimination the influence of the lightning, the structure would be destroyed. It is obviously, in that example, there is an energy factor, but a lack of the important component which to cause the birth and development of a living organism. How the organism would be animated, what and from where is the logic and, could be that logic a subject of knowledge? Or there is no logic, no algorithmic systems, all obeyed to the foggy natural force and evolution process, and, of course, to the magic properties of the matter particles?

All these questions are being asked not occasionally, they are essential in our desire to see life such as it is. Who should deny the next material let give first answer to these questions, all are keen to know it.

When talking about a living organism, let not skip the fact, the order of the particles of the matter, atoms and molecules in to a biological structure and animation are extremely complex, even for the simplest organism. To be an organism animated it needs a functional metabolism to be established, to build DNI and membrane that to keep all in one. That means a need of a complex device to create biological structuring, a device for the animation to be supported in any moment and in addition, a large memory.

From this point of reasoning we have to direct our attention to what are searching for. The Hypothesis denies the probability of self-creation and self-development sake of the mentioned above reasons. If someone is able to make a rational explanation let do it. We are going from here to the alternative for external intervention, intelligent design and spiritual genesis of life.

THE SPACE ENERGY AND INFORMATION FIELD

We have divided the physical fields in to three basic kinds. These, created (emitted) by the particles of inert matter and the third type, created by distinguished conditions. The fields of third type are of hyper high parameters and able to create closed stabile structures.

The simplest such a structure is a closed loop. The physical fields of third type are but more complex than a loop and being created as integrated coded structures or information streams. As we said, the conditions such a physical field to be produced is at the edge between the primary and secondary matter. We could suppose appropriate conditions to find in to plasma in the cores of the giant celestial objects.

The space is transpierced by energy and information field that translates large streams of information and integrated coded structures. The streams are directed to concrete objects in to the Universe. The common between all the energy and information structures in the environment of the ethereal media is the hyper high speed of spreading, the hyper high frequency, and hyper large volume of information available to be translated and memorized. Compared to these physical parameters, the ones of the fields, emitted by the secondary matter being limited by the speed of light are like a snail. An analogy is the comparison of the regular, called "snail mail" to the electronic mail.

The integrated coded structures are autonomous objects in the space which have independent existing and distinguished functional and structure features. Considering whether all of them are living structures, or some of them are only functional, auxiliary ones is out of the framework of validity of the hypothesis. The space information field and coded structures are conducted by the ethereal media, transpierced the matter of our Universe, besides, the size of a space object like the earth and the information related to it is not a problem to be operated and memorized.

Everywhere in the space intelligent coded structures are being spread. They make preliminary exploration and records of the conditions on concrete space object and areas of it in the context of probability life to be created. These coded structures are functional physical and chemical laboratories which research concrete environment, make comparative analyze, using their large databases, and by availability of appropriate conditions, start the process of some evolution.

These integrated code structures have features of structured physical field and they are able to action on the particles of the matter only on their components of semi condensed substance. As it was expounded in part one, a field is created by the emanation of the charged particles of inert matter, but is translated by the particles of the Media.

An electromagnetic field hyper of high frequency is unable to interact directly with the particles of matter because of frequency nonconformity. Respectively, the integrated code structures are able to act only to the zones semi condensed substance, being between the particles of condensed matter, such as protons and electrons, between the nucleus and electrons, the inter atomic and inter molecular spaces. These are the interface points to where the low energy space physical fields are able to action on the matter of our world. To the said nonconformity, the hyper high frequency levels of the physical fields of third type and the inertia of the secondary matter we owe the very slow process of evolution.

The functions and the problems to be solved by the intelligence functional hyper fields is to explore the physical parameters in a point, area and space, such as temperature, temperature changes, chemical analyze of stuff and gases. That analyze has an object to give exact answer to the question, whether in that point, area or space conditions life to be created are existing. By a principle, everywhere in the space, by presence of proper conditions, the creation of life is not missing, i.e., everywhere in the space the intelligence structures do their work.

In that part would not be considered evolution processes, this is a subject of the third part, "The Evolution". Here would mention only this, on the evolution process strongly influence the low energy of the physical fields which create the living biological structures – the organisms.

A subject of that part are the living biological organisms and the questions about their creation, development and existing. There are three basic conditions a living biological organism to be synthesized and animated. They are codes (algorithms), media in which the codes to be integrated into independent structures, matter and media in which the organisms to be able to live. It is important to be conformity between the integrated coded bodies and the biological ones. Further, the coded body will be called "controlling field" and would be distinguished from the biological body. In principle, the said coded body is an integrated physical field of a third type.

INTEGRATED ALGORITHMIC STRUCTURES

The personal computer, industrial controller, the cell phone and many other devices currently being in use by humans are based on algorithmic structures, integrated on hard material. They operate and memorize or translate information. By means of their electrical impulses computer and controller command periphery, machines, robots.

It is evidently, the biological organisms could not be operated by integrated on hard matter algorithmic systems. The contemporary official concept is, through the process of evolution the biological matter has created these the biological organisms by means of some "Natural force".

The Hypothesis claims, there are algorithmic systems integrated to physical fields of third type which are only able to command the biological processes. It refuses the presumption what is being invisible, undetectable it does not exist.

ANIMATION, MEDIA AND BIOLOGICAL SYNTHESIS

Currently the science and technology are able to provide extremely complex chemical and biochemical processes, as being a result of the development of the knowledge and technology. But, at the dawn of evolution the transition from inorganic to organic, followed by synthesis of simple and complex biological structures needs some force of movement.

Looking from the aspect of human's desire to create a living organism, the simplest ones are extremely complex. The most unpleasant is, each organism must be animated, not be only a structure of known biochemical substances. Besides, to be able to perform independent biochemical processes to support its structures. These processes are excellently explored by the science and are not a subject of that material. They should be mentioned as far as the main topic and direction of abstract and deductive reasoning require.

In that location the hypothesis is categorical – synthesis combined with animation of biological structures is available only by the influence of physical fields of third type. These fields only are able to provide forces to move the particles of matter. Having in mind the complexity to organize these particles in to 3D and functional structures, the fields must have the proper codes and physical parameters. In that case, the codes content the information about what to be done, and the physical parameters provide the possibility to be done.

As a consequence, the algorithmic system has to transform the particles of inorganic matter in to organic, to create biochemical compounds, structures, integrated in a simple organism. Considering that organism as being a matrix of particles, disregarding their concrete functionality, would be led to the conclusion, to order that matrix needs exactly definite and not simple logic. Besides, every one of the particles has to take its position in the outlines of a dynamic process. The general question is – is the matter able to do that being self-organized?

The Hypothesis introduces the concept of "Chaos and order of the evolution process". That means presence of unaccomplished stages of that process at every time. For example, if we could find creatures between the apes and humans, would consider that fact as a symptom of "evolutional chaos". If the living matter was spontaneously self-organized, it would be a permanent process and we should find intermediate phases of it during our presence on earth. Up to this moment we have heard legends about the "Snow men" and hunting of him. To capture a "Snow men", we should pierce him as being a worthy representative of the "evolutional chaos", even occasional.

According the hypothesis, only integrated algorithmic systems based on physical field of third type, at a high level of internal organization and parameters, with hyper high tact frequency could make all about we talk. Its pity, these fields could not to be indicated and explored, and of course, to be produced in the environment of our inert matter. But, they are being transited in the same media, where and our physical fields are. Sake of this, it was called "Universal Conducting Media"

FUNCTIONAL GROUPS OF CODES

All code structures are in their essence functional groups of codes, integrated in to code structures. The ability of these structures to exist independently and to be stable is crucial for the origin, development and support of biological life. The algorithmic systems integrated to physical fields of third type would divide in to two basic groups.

The first group is of these, being not engaged to biological organisms. To that group belong independent non material living creatures, intelligence and explorers and supposing, many more. The second group comprises the ones, integrated to living biological organisms.

In the spot of our attention fall the species of the second group basically. For the ones of the first group we could only to suppose, but have and some facts that would be explained further in the material.

The Hypothesis follows the logic the visible part of an organism, its biological essence is being structured as reflection the invisible, elusive part or the physical field of third type. Because the functionality of the biological part is being explored at atomic – molecular level, there is no meaning to repeat what is well known, translating it to the algorithmic structures of the field. Independently of that, would need some particular repetitions to make the hypothesis clearer. The basic claim is, to each biological organism is being integrated one algorithmic structure that belongs only to that organism.

FUNCTIONS OF THE ALGORITHMIC SYSTEMS

The hypothesis claims no complex organic compound structure and living organism could be created by the spontaneous self-assembling of material particles. From here it follows an operating and controlling facility is needed to possess these functions. This is a physical field that belongs to each living organism from the moment of its birth up to the end of the material life.

To ease the explanation further will distinguish the material organism calling it organism and the physical field that belongs to this organism calling it "the field". We have in mind this physical field is a hyper complex algorithmic system possessing all codes needed the organism to be operated and controlled during its material life. The hypothesis considers all material structures of the organism as mirror of structures of the elusive field.

The field operates the vital functions and behavior of each living organism. The field as any electromagnetic field is property of the Media. But there is a lack of energy in the ethereal media, respectively, the field experiences a deficit of energy, therefore, it actions to utilize maximally the energy of the material world.

The energy of the field and the energy of the material world are distinguished essentially by their parameters. The electromagnetic waves of the organism's field are of hyper high parameters, foremost frequency, and unavailable within the material environment. But, this makes it able to realize hyper quick reactions, large memory and to control simultaneously many complex processes.

To imagine how complex are these processes, and which information streams are translated to be analyzed, let make a considering as follows.

The eye receives in every moment the information that corresponds to the visible scene. The number of points of vision is approaching to unlimited. This is one of the conditions to obtain realistic reflection, imagination of the reality. Each of these points brings precision color, light and 3D information. In addition, the information could be extremely dynamic.

By the same conditions, an optical apparatus, regardless how complex and perfect is makes the scene represented by pixels, and by relative dynamic motion between the optic lens and scene, the picture is going to be distorted and of a slur. It is problem to acquire, transfer and memorize the information of a series of pictures by a motion of the optical lens towards the scene, and to take clear, fidelity pictures. There are problems to focus by dynamic mode. Compared to this, the optical system of a living organism transfers much more information. Simultaneously with the optical, the organism receives acoustic, chemical (smell), physical (temperature, damp, air conditions, wind, whiff, etc.). To this information to add the actual information that corresponds to the internal organic status, what we have called a matrix of information. In conclusion would be said, the organism processes large streams of information.

INFORMATION MATRIXES

At that point of our theory we would define two basic matrixes of information being received and processed by the organism. The first matrix would call "internal matrix of information", and the second – "external matrix of information". It is out of doubt; the streams of information from both the matrixes are extremely consistent and dynamic, being a subject of change in every moment.

Let follow these consistent, dynamic streams. From the sensors they pass through the peripheral and central nervous systems, and being distributed, approach to the corresponded section of the brain. Further, the information is being compared to the one memorized in the database and a process of comparative analyze is going. The comparative analyze produces reaction that could be emotional or a command. The large streams of information need respectively a large database for comparative analyzes, the large database requires a large memory. The content of the database concerns information about colors, textures, smells and aromas, acoustics, knowledge, experience, scenes and many more. As a result, very quick and relevant reactions have to be produced.

According the contemporary concepts, the information is being stopped to the brain where it is processed. Here we talk about the vertebrates and the human.

BIOLOGICAL STRUCTURES AND ELUSIVE FIELD

The basic problem before the Hypothesis is to find arguments, supporting points for its claims. The matter does not allow direct proof, and the arguments, if are proper and relevant, and well described could make claims more available and realistic. Regardless, direct proof is unavailable, further in the material would be narrated and expounded a case that could be considered as being indirect evidence. The basic problem is: do the algorithmic systems really exist? This is the claim and it contradicts to the contemporary materialistic concepts of biological, pure material evolution and development. The supporting points are as follows.

First, there is no evidence the particles of the matter to do anything except what the physical laws and regulations would induce them to do. If the origin of life and the evolution are of material or biological character, it is not designed but spontaneous. When evolution is not planned, it is based on chaotic and permanent transformation of the species. The third point is the complexity of the biological structures, a subject of these lines. Really, the biological structures are developed to extremely complex levels. They have created hyper complex internal structures, morphology, and behavior.

The hypothesis directly claims and this is essential for it, the biological structures are created by means of external influence, the physical fields of third type exist and without them biological structures are unavailable. From here the second problem follows before the hypothesis. Considering the algorithmic systems - physical fields to exist we are unable to explore them. What can to do is to try making in a logical way a maximal approach to their construction and relation to the biological organisms.

There are many sources of literature, where the elusive fields are described in details, without hesitation. This is too boldly for an area, where nothing is visible, sensible, explore able. To the hypothesis this way is not allowed. What we can do here is only step by step to accept or refuse arguments, facts, considering.

It is out of doubt; the biological structures are able to create a wide range of complex functionality. Our problem is to decide, where the biology finishes and where the elusive body begins. We have followed the path of the information from both the matrices – the external and internal from the sensors through the nervous system and stopped to the biological brain. All in that path is biological, including the cells of sensitiveness. These cells are micro chemical laboratories, able to distinguish a large number of smells and aromas and different feelings.

The Hypothesis claims, the information does not stop to the biological brain. It continues from it through the gates to the corresponded zones of the elusive body – the field. What part of the information is being processed and memorized in the brain, and what part – by the field? The contemporary theories consider the brain as being the only information processing organ of the body. There is no problem the biological structures to be enough complex to do that. But, there is a slight restriction, the tact frequency of the command and processing impulses. The tact frequency, realized by the inertial biological structures never could process the large stream of information from both the matrices. The reason is this is inert material structure which processes very slow. These structures are restricted by slow tact frequency defined by the speed of light. They are unable to create the large information database and the required memory.

The hypothesis claims, only the field with its hyper high tact frequency is able to control the processes and to realize the large memory. The brain is operating the organism at the biological level, but actions and as an interface between both the bodies – the biological and the integrated to it physical field, the elusive body.

The field has two levels of integrity. The first level is the self-integrity and the second – integrity to the biological body. Once, it is independent and twice – closely bound up to the biological body for the term of its material life.

There are regulations and even laws that constitute the functions and behavior of the fields. During life of the biological organism, it is inseparable from its field. Both distinguished sides of one and the same definite body exist in to two quite different Medias.

Definite body of a human is being featured as his unique "I", personality, identity. As was boldly emphasized, our material media is being immersed into ethereal media. The ethereal media is universal conductor for all the physical fields, including for the fields of organisms, as well and this makes their self-integrity available.

The Media does not allow observation and exploration of the physical fields of third type from position of our material world. It is known, the field closed to some organic structures was photographed as a vague image, and the energy of that field has been detected (Kirlian photography).

COMMUNICATIONS

There are two basic directions of communications of the ethereal bodies. The first is towards the ethereal media, through which gate the field exchanges information with the Space Information and Energy Field. The second gate or system of gates exchanges information from and to the biological body – the external and internal matrices through the brain and biological systems.

Basic regulation of the behavior and feature of the ethereal bodies is related to their communications. Direct communications between both the media are not allowed. Such an organization is not accidental.

The field during the term of life of the biological organism integrated to it is directed to the material world with one basic design – to reproduce, develop and support life in the media and environment of the material world. Each field, being directed to that media, because of its lack of inertia by processing the information, makes to the organism strong and sufficiency feeling of belonging to the material world, and never a feeling or sense for the elusive, with some exceptions which we call "paranormal".

The field is protected from the side of the external information field (the Media), so that penetration and disturbances from that direction to be reduced to minimum. This is what we call "Fire wall". The fields of the organisms are distinguished in the same way, as the organisms. As well, they have different level of protection from the media where fields exist (the parallel world).

There are two basic types of influence from the parallel world. The first is intervention from the space field, besides the protection of the individual field becomes locally broken and there penetrate external information or influence. These information and influence could change the structure of a field as well its content. The second type of influence is a result of interference between given field and the fields of other organisms. Let not forget, from the ethereal media the streams of information float without inertia, with hyper high speed, invisibly for us, penetrate the matter of our world and are not being influenced by it.

ENERGY AND STRUCTURE

It is considered to talk about positive and negative energy, when the question is about the fields and their status. Physically, the energy is not positive or negative, it is simply energy, exists or not and possessing defined physical parameters. Positive or negative sign could be attached to energy streams and relatively or conditionally. For given physical field positive is stream of energy which is being charged, and negative the one that is exhausted or withdrawn. The energy exchange is strongly obeyed to the physical laws and regulations, besides, interference between physical field of third type and media, or other field may cause origin of positive or negative energy streams. At the first case, the body is being charged, and at the second – discharged. Sake of the low energy level of the processes by the algorithmically structured fields integrated to material organisms, a field is able to interact with the matter, in particular, on biological structures by means of fine, low energy impulses of hyper high frequency. There are two basic factors for the ethereal bodies to function properly, structure and energy.

STANDARD STRUCTURE

The standard structure is an ideal structure of a field integrated to defined biological organism by which all code groups and memory are being ordered properly and this corresponds to one perfectly functioning organism.

The standard structure of a field integrated to defined organism of given specie is the formula of its creation.

The structure of a field integrated to material organism is designed as hierarchy of code sub structures realizing definite functional directions. It is set by forming of organism's embryo. But, for distinguished species, the primary set of basic structures has been done on the corresponded stage when that species had been created.

That basic specimen of the field we would call "basic formula" and it contents the basic information related to the species. For example, for human that information is related to his proportions, organs, systems, mentality, awareness, personal identity etc. The basic formula of controlling field is being defined and implemented at a moment when the species had be originated. That topic is a subject of the third part of the book, "The Evolution".

The basic formula for defined species is analogical to its structure at the original conditions of creation. Later, in the process of development that basic formula is not a subject of change, it is preserved, but tolerance of changes is being preliminary coded in to it. The basic formula or the previously implemented codes would call as well "Standard". The standard includes the basic features of the species, making it distinguished than others. The standard for human includes height, proportions of body, head, limbs, bipedal, manner of moving, pose, all external features, balance status, internal organs and systems, and foremost, the mentality and personal definition.

Basic regulation of the behavior of each organism is the aim to approach to the standard formula. That means, first the aim is in the process of development of species, and second – to return back to standard when conditions of life have changed the species.

The preliminary coded probability to change out from Standard would call "Tolerance" in which the species could be modified, or to mutate during the term of their existence. That tolerance is related to the future species' life, eventual and probable change of the environment, or restricted development in some direction. Preliminary coded probability to change (mutation) does not influence the basic standard algorithms; it is foreseen and implemented simultaneously with the standard. In that context, modification of the standard means appearance of new species.

The standard for a human could find in art works from ancient and contemporary authors – sculptures and paintings. Erected body, limbs and external features which we accept as being exquisite, all these features have been coded at once by human's design. Along with this has been foreseen, human must be modified during his development. His codes of modification are related not only with change of living conditions, but foremost to human's mentality development.

By the animal species the previously coded tolerance for modification are closely related to the aim to survive, as well, to environmental changes. This is the explanation of the protective coloring, perfectly corresponded to the surrounded scene. Representatives of one species that live in distinguished geographical latitudes have different colorations. The fish has light colored tummy, because the light is coming from above, and a dark or properly colored upper side, because the bottom of the basin is dark or colored.

It is advisable, to spend some time and to think of, how this is possible to be self-created. The logic of that adaptation is too complex, the physical performance – more. That process needs precision analyze of environment and concrete background, next the systems of the organism to be set in action to change the features in accordance to that analyze and to harmony to the background. The same is about adaptation to all the physical conditions, such as temperature, changes of temperature, winds and more.

From a view point of the physics, each organism's life is complex, and for its support large information must be processed. All diversions are included in the original model of the field which corresponds to standard living organism, but they are applied without breaking the standard formula. Therefore, when the diversions are temporary, after elimination of their agent, the organism aims to return to its original status that corresponds to the standard view, accounting new environmental changes if such are appearing.

Relating this to the human, during his existence a lot of diversions from his standard model are becoming necessitate. They are a consequence of the chronology stage of development, model of life, work, living standard, food and many more. These variable conditions could affect causing diversions from human's original formula, such as change of skin color, hair, height, limbs curvature, sharpened or bluntness of the senses etc. In all cases available, after canceling the reason of diversion, the controlling field follows the original codes and commands the biological one to return to the original. This is a process with inertia, sake of the inertial biological processes and never could be accomplished during the existence of one generation.

An important group of diversions from the standard algorithms of the fields are the irrevocable ones. For irrevocable diversions there are no preliminary tolerances of the standard and they are not foreseen. To this group belong changes in the environmental and living conditions that have been not expected. Such are artificial electromagnetic fields, radiation, deceases and conditions caused deformations or destructions. Sake of lack of before defined and implemented algorithms, functionality and low energy, the fields are unable to manage the biological organism to return to the standard. The only way such problems to be overcome are the human's mind, science, experience.

The structure of the fields is closely related to their functionality and resource of energy. The field does not dispose of high energy levels and manages the belonging biological organism by means of fine electromagnetic impulses. The structure of the field being closed to its functionality hypothetically contents functional areas, namely, functional part, memory, communications and protecting part.

The numerous species are classified by their defined features and they make differentiated groups in the hierarchy system. The functional part of each field corresponds to that of the biological organism which operates. In accordance to this simplified classification, the fields are simple, root – trunk and brain – vertebral. To the first group belong the simple organisms, to the second – the plants and to the third – vertebral animals and human. This is a simplified classification for the purposes of the hypothesis only.

All living organisms have operating fields. The basic function of a field is creation and reproduction of species in the framework of the process of development, animation of living organism, supporting of life in the definite environmental conditions.

Creation of each subsequent during the development of the species is accomplished by means of intervention and change of the field of the preceding organism. The preceding organism after the intervention begins to follow the algorithms of the modified field what is an inertial and slow elapsed process of thousands and millions of years. The basic reasons are low energy levels of the controlling fields, respectively, low energy processing impulses to make changes of biological structures and the high level of inertia of the biological mass.

ENERGY CONSUMPTION AND SAVING

Energy is the most important physical parameter which determines the ability of the operating - controlling field to command processes of the biological body to function normally. Both, the field and the organism exist within distinguished in principle media and therefore have different energy parameters and levels. The environment of the material world is basic energy producer and the field could only to utilize it after transformation. The lack of energy in the ethereal media leads to need of its economical utilization.

The sustainability of living functions of an organism is related to processing of information streams of both the basic information matrices – of the organism and environment. The basic aim of the field is, to process the organism by minimal energy consumption. That means, they aim to maximal utilization of the biological structures and their much higher level of energy, and to do that with minimal spend of own energy.

The creation of unlimited number of species is a reflection of the need to save energy to operate and support the organisms. The extraction of readymade biomasses to feed organisms is much economical and rational than synthesis of chemical elements. Instead of that synthesis, the biological organisms simply extract complex compounds useful for their existence. Thus, the processes are simplified and the energy consumption from the ethereal media reduced. To utilize the energy of the biological organism, the biological structures are developed to high levels. A biological organism plays and role of amplifier of the week impulses generated by the field, conducting them to the corresponded zones of functionality.

What we call today Eco chain is reflection of this economical and rational creation of conditions for the organisms to exist. Within the material world the organisms need material substance to sustain their life – this is food. But food is created by organisms – their factories fulfill this plan and realize the system of rationality.

REACTIONS, EMOTIONS AND FEELINGS

The organism's controlling field forms all senses, emotions, and reactions, current and instant processes which are being processed within the organism and as reactions to external factors. All complex reactions, such as feelings, emotions, senses, etc. are coded within the structures of the field. Only hyper high frequency and fine impulses based on hyper large memory provided with database for comparative analyze make these complex processes available. The hypothesis keeps in mind the probability of biological structures to be organized to extremely high levels, but rejects the probability of their self-creation, without controlling fields. The role of the biological structures is restricted to creation of all sensors, conducting of signals, impulses, interface to the controlling field. Generally, it is a problem to define where exactly ends the biological functionality, and where the field begins. We are able only hypothetically to try to determine the areas where the biological structures are being unable to act, the complex reactions, emotions, feelings, where the complexity of the phenomena and inertia of the matter are factors of restriction. The assertion of the hypothesis at that location is, the biological structures are not created, but operated and controlled by their fields. The hierarchy is: Field of organism; field of a cell; biological structures.

REPRODUCTION OF ORGANISMS

We would direct our attention to the humans and their reproduction process within the context of the hypothesis. Here a reproduction we would consider as conceiving of a new organism, able to begin its development. We introduced an information matrix of organism, being integrity of all particles as atoms and molecules, but organized in to micro and macro biological structures – cells and DNA, hormones, organs and systems. The information from that matrix is conducted to the brain and from there – to the field. During the process of development of an organism all these components of the organic matrix must be changed in absolute synchrony, taking their right places in function of the time.

The field as a virtual subject possesses a function of auto copy. By defined conditions it is able to copy all or part of its codes, representing its features in within a new but identical structure. Copying process is going by strictly defined regulations, originally implemented in the algorithmic system of all controlling fields.

First, the process of copying is going ever in exactly defined moment in the framework of a reproduction process. The copied field is always attached to a biological reproductive structure. This is the male or female sex cell – respectively spermatozoid and ovum. The copied controlling field and the biological reproductive structure are being created simultaneously in a strict synchrony. In that way, every reproductive body, such as ovum, spermatozoid, seed and their structures are created by attachment of controlling field which is integrated to them from the moment of their creation. The field transfers the specific features of the parent's, and all information needed for the new development.

The copied controlling fields, attached to biological structures by reproduction of unisexual organisms carry all information to the new one. The copied fields of bisexual organisms do not carry the whole information needed to origin new one. There are distinguished two information carriers, attached respectively to the ovum and spermatozoid, which are unable to origin organism before fertilization.

By ovulation, monthly, the mother's field commands creation of biological reproduction cell – ovum, and simultaneously copies, creating new controlling field, automatically integrated to the biological (to ovum). Life of ovum's controlling field lasts so long as life of the ovum.

The father's field commands simultaneously creation of biological reproductive cells – spermatozoids and integrates a copy of itself to every of them. Life of spermatozoid's controlling field lasts relatively no long after ejaculation.

As result, each ovum and each spermatozoid are provided by own elusive field. By fertilization these fields create a compound controlling field of the originated organism being is able to develop it. This common field carries a combination of the features of the donors inherited. All the information is recorded and within the biological structures. If the ovum has not been fertilized, all the fields of all reproductive bodies are being disintegrated and destroyed.

The information, carried by the field of the originated organism is much more, than the biological information, copied to the organism. It consists not only of inherited features, but of all needed to command the organism's future existence and development. In the term of uterine development, the influence of the mother's field is essential, because of existed biological ligaments. The originated field begins to command all processes of the new organism. It carries and the basic, standard formula by which it develops the organism strictly following the complex algorithms implemented.

DEVELOPMENT OF ORGANISMS

Physically, development of an organism means strictly arrangement of all its particles in 3D positions and in the time. In every instant the organism is changed, besides its particles – atoms, molecules and the built by them biological structures are being arranged and rearranged in a harmonic order one to other. To return to the basic problem of the hypothesis would ask again the question – if they do that being not influenced, what tool should motivate them?

The Hypothesis is categorical – the hyper high tact frequency, hyper large memory, hyper complex algorithms of their controlling integrated fields (physical fields of third type). Thy sent simultaneously billions of low energy, non-inert precise impulses which are the tool that motivates the processes of development.

TIMER FUNCTION

Each field which commands a biological organism has precision timer measuring micro parts of time, hours, days, months and years of the earth's life. It starts to function from the moment of the conceiving of each organism and measures time during its whole life.

The development of the organisms is function of and harmonizes to the time of our planet. It begins from the moment of fertilization, continues with uterine development, growth, mature age, ageing. Each phase of development requires precision arrangement of the particles in to biological structures. The phases of uterine development and growth require in each moment new particles to be added to the previously constructed. That means a complex dynamical timing process.

To achieve that, the fields are provided with algorithms for precise timing. They exactly correspond to the earth's cycles of year, month, day and night timing. All processes of organisms are synchronized with these algorithms. Each biological particle, organ and system must grow and to be arranged so that the proportionality and balance of the whole body, the internal matrix and its functions not to be disordered during the time.

The existence and all phases of life are being put in to frames and correspond to the General Harmony of Life on Earth. There is one parameter, distinguished for the species, the average expectance of life. For the humans it is limited to 120 earth years. That means, by best conditions of life to reach to that value. The real life is a continuous train of stress, polluted environment, food and drink that do not meet the requirements, sedentary or heavy life etc. All that puts the average expectance of life deep bellow the coded value.

The timer of the field of each species measures not only the time in years, months and days, but makes precision accounting of all processing relatively the duration of its life. The timer of the fields distinguishes plants that live by conditions of four seasons from these that live in polar or equatorial areas. It distinguishes the animals that need short time to become to matured age from humans that need long years for that and that time as a specific fraction of their average expectance of life. For example, animals like dogs and cats that live up to 15 years grow and reach matured age in their second year, and humans that need 15 - 25 years to get knowledge, skills, experience.

AGEING

The average expectance of life for humans is preliminary coded within their standard algorithms. As it was mentioned, it is limited up to 120 years by ideal conditions of life. Every diversion from the ideal conditions of life leads to decrease of that value.

There are two basic reasons for aging. The first is algorithmic and is performed by the timer of life. It measures the time from the moment of birth, accounting continuously the time remaining to the value of standard algorithmic duration of life for each species. The second is physical and is based on restrictions, being set during life. Foremost, this is insufficiency of energy and structural changes. The problem of aging is considered as being very important and would be expounded according the hypothesis's reasoning.

When processing the information from the matrices, the external and internal ones, creating reactions, followed by forming of command impulses back to the internal matrix to organism, the controlling field exhausts its energy. During the time, each organism is object of influences, such as, diseases, stress, polluted environment, other organisms etc. The controlling field receives additional information from the matrices, foremost, from the internal, that corresponds to these influences. Respectively, it must form additional reaction, mean, additional number of command impulses to eliminate the consequences of the given influence for the organism. To do that, the controlling field spends additional quantity of energy. Each influence could be perfectly eliminated, but could cause a slight, small or essential defect in the biological organism. The number of defects rises during the time. For and young organism it is not essential, and the controlling field of it is able to eliminate influences being not compelled to spend a lot of energy.

It is not the same by an aged organism. The number of internal biological defects has been increased in a progression, and the energy to eliminate new influences is respectively raised. The controlling field is becoming unable to recharge energy effectively. In that case controlling field starts its algorithms of emergency. It begins to reduce the information from the matrices, to process reduced quantity of information and as result to feed back the biological organism with reduced number of command impulses. That way, it ignores influences, does not eliminate consequences of them and as result, the number of internal defects rises.

All this is a consequence of necessity to keep the energy consumption equal to the level of available recharge. The emergency codes of controlling field reduce the density of information from and to both the matrices to provide that. Reduced information means worse sensitiveness, including vision, hearing, and loss of ability to resist against influences. The controlling field is no more able to arrange the particles of biological organism to their original places. The organs and systems become to be deformed. Controlling field neglects the command of metabolism and other function, in its aim to save energy. All those are the symptoms of becoming old.

In that context are included and the cases, when a biological organism is being defected by the time of its birth, or becomes defected during life, besides the defect is essential and ineluctable. In all cases like these, the controlling field needs a lot of energy day after day, to support life and early become exhausted.

In that context we should find one more (indirect) argue for our theory. Talking about aging, we mean all organs neglected by controlling field because of insufficient resource, or obeyed to the algorithms that limit material life. But, there are two biological bodies that never are being affected by ageing and never neglected by controlling field. These are the cells responsible for reproduction – spermatozoids and ovum. We are witnesses of plenty of cases, even of public distinguished individuals that being advanced in years have absolutely normal generation. The appearance of old people and their organs are changed, but their reproductive cells not! There are no old spermatozoids and ovum, they simply are or are not. This is because of the algorithms that do not permit the organism to produce deformed from ageing reproductive cells. Controlling field neglects all others, but never these; it has special codes to do that. In case, when resource of energy is totaling insufficient, it simply stops to produce reproductive bodies, but never ones with neglected quality.

During life there is one more factor that determines the processes described above, the structure of the physical field integrated to given organism. The role of that factor is a subject of the next topic.

RELATIONS CONTROLLING FIELD ORGANISM

There are three basic factors that feature the status of a field and its relations to the biological organism. They are structure, energy and communications.

The controlling field has an ideal structure which could be disturbed by various influences. It consists of all code structures in their integrity that construct the personality of a human. These code structures define the state of the organism, individual identity, personal and social mentality. From aspect of psychology, they build human's consciousness and sub consciousness. The latter have mostly abstract character, but they command the organism's behavior.

Regardless consciousness and sub consciousness are abstract notions we would relate them to supreme algorithmic areas of the field structures. These algorithms command the conscious actions of an organism, and by human – of his personality. The personal behavior depends on these areas. By availability of some destruction of this algorithmic system commence diversions of individual behavior as a direct reflection. The consciousness is that supreme structure which commands the actual behavior of an individual. The area of structures that corresponds to consciousness is external, closed to the information exchange of both the matrices. Therefore, it is more vulnerable to external influences which change the structures, respectively, the algorithms of individual identity and behavior.

In a difference, the sub consciousness is realized by internal code structure, being protected by direct external influences. It is closed to the memory from where is able to acquire information that could be called “sleepy”. In that way, the code area of sub consciousness is able to alert the external, action area of consciousness and to relax the broken structures. This is a permanent or discrete process, but by more serious destructions and more serious procedures are needed.

The communications of the controlling field are directed to two basic locations. The first is towards the ethereal media, mean, the parallel world of hyper high frequency and connects it to the Space Information and Energy Field. The second basic communication is directed to the biological organs and systems of the organism and the material world.

RELAXATION OF THE CONTROLLING FIELD

During its existence the controlling field is under attack of the space hyper field to and from which should transfer in both directions energy and information. Second serious influence is a result of interference with other controlling fields, integrated to other biological organisms. By closed contact between material organisms the integrated to them fields interfere, but do not lose their integrity, that is physically possible. Problems are being caused by mutual influence, such as energy charge – discharge, loss of information and destruction. Meeting with another's field corresponds to define emotion that could be sensed by organism and to be a pleasant or unpleasant feeling. The interference with fields of some people, plants and animals could relax the controlling field.

To be able to function normally, the controlling field must be put in ability to relax its broken or disturbed structures and to charge its exhausted energy to maintain the integrated organism. There are two basic relaxing processes related to the biological organism and the integrated to it controlling field. The first corresponds to the material body; it is metabolism. It is explored to high average. The second is the structural and energy relaxes of the controlling field.

During the processing, the controlling field is closely connected to the biological organism. Communication gates between organism - field are at status "open". The energy consumption is maximal, structural disturbances – probable. The controlling field loses energy to process information and to send fine impulses and in a moment gives indications for that. The indication is a call for the biological organism to sleep. Normally, by absence of extraordinary reasons, the controlling field follows the cycle day – night, being able to charge enough energy and to restructure during the night to be active during day.

The controlling field needs to abdicate out from its responsibilities to the integrated organism and to become free to relax. It breaks most of the communications to the integrated organism, supporting only the functions being crucial for its existence, like heart pumping, breathing, and metabolism. The processes which need command impulses are strongly reduced. Information from the external matrix is minimized to zero, all sensors are closed. Information from internal matrix is also reduced to low levels. Only by these conditions the controlling field is able to relax. Deeper sleep, better relaxation processes.

During its relaxation process, controlling field withdraws from the biological and translates closely towards ethereal media. During the process its algorithms for restructuring are activated. It is more closed to ethereal media than to the biological body, to the parallel world of hyper high frequency than to the material world. Therefore, it is able not only to charge energy, but and to receive information. Keeping in mind, the controlling field and processes in ethereal media are non-inert. The return of the field to restore its communications to the organism and its brain is non-inert as well. A slow return from a sleepy to alert may be caused by exhausted and still not relaxed field which does not like to take solution to return, or slow inert biological processes.

During the term of relaxation the controlling field leaves between itself and the biological organism these communications only which are crucial for the organism. Such are support of living functions at level of metabolism. In that way the field minimizes its functions to the least energy and functional levels available. That allows the field to include its algorithms to restore structures and to charge energy. In that term the field has no communications to the material world via the organism's sensors because they are eliminated. In the same moment its communications to the Earth and Space Information and Energy Fields are enhanced.

The memorized original, standard algorithms allow restoring the original structures, removing defects, repairing broken parts. The process depends on the level and deepness of these defects and the ability of the field to be restructured perfectly or not.

The lack of or insufficient energy makes operating the integrated biological organism difficult or impossible. The controlling field reduces the information streams from and to both the matrices and gives indications for its exhausting. To illustrate this would refer to reproductive processes by humans. Copying of mother's, respectively, father's controlling field by reproductive process is accompanied by increased energy consumption. But, it is going distinguished for men and woman. Ovulation, respectively, copy of woman's controlling field is once per month. By man, copying is performed by every sexual act. The controlling field of men during ejaculation loses a lot of energy to be copied and gives instantly indication – he is getting sleepy. The young organism could donate more energy because of its high effectiveness. Compared to men, the woman's reaction is quite different, because she does not lose energy to copy her controlling field during sexual act.

EARTH ENERGY AND INFORMATION FIELD

Charge of energy is crucial for the controlling field to function during the next active period. The impulses which operate the organism are of low energy, but their integrity during active period exhausts the field. The physical parameters of that energy do not coincide to anything of the material world. They are distinguished foremost by frequency of electromagnetic waves. The field could be charged only with hyper high frequency energy which could not to obtain directly from fields of material world. There are two basic ways that energy to be supplied.

First is when from Space Energy and Information Field to given controlling field a stream of concentrated energy is being directed. The organism that belongs to such a field relaxes quickly by relatively short duration of sleep. To achieve that, the Earth's physical Field is being pierced by the Space one and functions as local field within the Universe.

The Earth's Field is charged also by energy of the material world. Such energy exists in abundance but must be transformed so that to approach the physical parameters of the hyper fields. Transformers of energy parameters are interface structures, special functional fields, or such integrated to organisms. The energy balance of these interface structures allows to produce a surplus of energy, transforming that of the material world (of sun and earth's core). That amount could be directed to elusive bodies integrated to organisms. There are known fields of trees like horse-chestnut and other plants, animals that charge with energy controlling fields of human, relax them and in some cases help injures or diseases to be cured.

It is logically considering specialized structures which transform energy from the existing sources of the material world. These structures are crucial for our life. Emitted of them energy at defined points and zones of the earth's surface or deep into it corresponds to optimum conditions of life. It is known; animals are sensible to such energy sources and make den selection carefully to be closed to them.

ELUSIVE BODY, ORGANISM, HEALTH

Defining Standard for human, respectively, standard human's controlling field, ought to have in mind the numerous factors which form an individual. The combination of these factors leads to the unique of each one. The train of reasoning leads as well to the absolutely impossibility controlling field to arrange particles of integrated biological ones in identical way. Each individual has at some level diversion from the standard formula. That diversion is preliminary foreseen in the standard codes, but in some cases could extend out of it.

To first case are related diversions of size, external features and internal characteristics. These signs are formed at first by heredity and second, during the existence of organism. To the second case are related all diversions being not preliminary expected in the standard formula, such as heavy diseases, injures, inherited faults. Considering the energy balance of controlling field as being distinguished factor for its and of the integrated to it organism vitality let have in mind diversions that have or not previously given algorithms to restore (or return to) standard that should correspond to affection or aging.

Operating the biological organism, controlling field permanently processes streams of information from both the matrices. That is performed fully during the active term when all senses are alert and partially during the relaxing term, normally following day and night cycle. The controlling field makes comparative analyze of information regarding the standard algorithms memorized in its database and forms reactions. Talking about organism we have in mind foremost the matrix of internal information.

When the external conditions influence the internal status of an organism, the information streams from both matrices interfere. Reaction of controlling field to diversions having been previously foreseen in the standard formula is directed to eliminate the problem. In these orders are being included immune system, blood coagulation, change of body's temperature etc. For cases which are not foreseen in the algorithmic system and direct reaction is unavailable, by humans is given opportunity to solve problems his reason setting in action.

As far as we do not talk about a reason by animals, but for a lower level of forming of behavior, their algorithms are not much simpler. In some cases they have more sensitiveness, foremost, to gravity and magnetic fields of the earth. That allows them to orientate and perform actions being impossible for humans. Their actions are previously coded, their algorithmic system performs comparative analyze of the physical parameters and commands following actions. Such actions are migrations with a purpose to survive or reproduction. They are not based on reason, but on automatism realized on complex algorithms and repeating an action in a similar way.

The reason by humans makes things quite distinguished. That is a topic, out of the hypothesis because it is well researched and known.

The group of diversions, not foreseen in standard formula is related to such that could not be analyzed and respectively, to perform reaction to return to the standard is problematic or impossible. This group contents diversion from weigh, shape, proportions. Organism is going out of control by integrated to it controlling field which is being compelled to arrange the biological particles at positions extended out of standard. That causes problems of the controlling field, especially increased energy consumption. By such diversions if the field is unable to support energy balance, it requires more time to relax, reducing the active term and efficiency of the organism. Simultaneously, the biological processes require more time, becoming more inert, and as a consequence a need of more energy becomes.

The aging was expounded, but here would make a brief summary in the context of the topic. The changes, related to aging are previously defined in standard formula, but restricted in a framework. During life, diversions from standard are being collected and lead to defects in the biological organism. The need of additional energy rises in a progression during the continuance of life. Within the standard formula such development, when the amount of energy is unavailable is previously foreseen.

The algorithm for that case dictates the insufficiency of energy to be overcome by strictly defined way. The energy misbalance is compensated by reducing the information streams and neglecting some of the functions. The controlling field is unable to arrange the particles of organism at their 3D positions as perfectly as needed for to support their optimal functionality. The eye balls being not supported at their nominal positions are unable to focus, and many more irrevocable changes are well known and should not be a subject of that theme.

When some individual has a good, strong reason he is able to lead his organism to stay closed to standard, minimizing the number of defects, respectively, the need of additional energy. The care for organism when is literate and systematic could lead to support for long years the density of information streams from and to both matrices at level to prevent early aging. When say a strong spirit to keep the body, there is a question namely about the human's physical field, which optimal and strong structures to function properly and to operate correctly the biological body.

Finally, in an instant, the organism terminates its material life. That occurs when controlling field is becoming no more able to operate it. That status is a result of the permanent comparative analyze of the stream of information from the internal matrix when it shows that.

There are two basic reasons – loss of functionality of basic organ and lack of energy. Controlling field is the first instance that takes solution to leave the biological organism. If, add of energy is possible and it is sufficient to be rehabilitated, or appraisal that given organism must live more and has opportunity to do that, the controlling field returns back, for example by some cases of coma. In cases of irrevocability, controlling field leaves the biological, breaking all the communications to it. The biological organism is no more able to exist and transforms in to decayed biological mass.

OPERATING OF AN ORGANISM

As we were convinced (or not), each organism needs operation to support its existence. Both the matrices, the external and internal in every instant translate large streams of information and the organism needs a large number of command impulses to function. Closing eyes, we are able to consider, the matter is endowed with reason and able to be self-organized in to complex structures. But, when open eyes, should see, there are no indications the matter and its particles to be in possession of such qualifications. The particles of our material world simply follow the laws and regulations of physics, which dictate their behavior. Or, as it was formulated in the first part, "Gravitation", the behavior of the particles of the matter is defined by the conditions in to which they are being settled. Even, a long lasting evolution process would change nothing of this.

The controlling fields exist in media that defines their possibility to be integrated and to function with hyper high physical parameters. The lack of energy in that media determines the aim rational and economical structures to be created.

The functionality of an organism requires differentiation of functional organs and systems. The field does not process a number of undefined information points, but differentiate functional organs. In that way, the stream of information from and to internal matrix respectively, energy consumption is reduced by increased effectiveness of function. In our review we do not distinguish the organs and systems as being functional bodies, but like micro (cells, hormones, genomes) and macro bodies. We are not interesting about their concrete functionality, considering them as 3D objects of particles, with exact position, shape, and size. Each organ sends information to controlling field and receives processing impulses from it.

A rational and effective organization requires construction of complex biological structures that to be able to perform as more as possible functions to reduce the number of command impulses from the physical field. In that way, the field saves energy, leaving more of its functionality to the biological structures which do not experience a deficit of it.

The controlling field commands each organism, arranging through its algorithms atoms and molecules of chemical elements and compounds in to structures of organic and living matter. It creates complex biological structures that to serve all functions related to the existence of the organism. Basic of these structures is the cell, supreme – the brain. Brain is basic mediator and interface between material and ethereal structures.

Signals from both the matrices are permanently transferred via conducting structures of peripheral and central nervous systems. Through the gates of brain they continue towards the elusive part – the controlling field which is mostly concentrated and energized in the brain zone (fontanel). The brain is able to conduct impulses between both the Medias. That is performed by external for the brain interface formations being at the boundary between both Medias. These formations are ethereal parts of the particles of biological structures, therefore, universal and able to transform energy.

Signals of optical, acoustic, of all peripheral sensors, of internal matrix are being conducted to these points of communication as biologically realized current with inherent electromagnetic field. Their physical parameters correspond to these of the material media. At the end points of the brain they are perceived by the interface structures of the controlling field that are closed to the brain, transformed by frequency and transited to corresponded divisions for comparative analyze.

Back signals, result of the activity of the controlling field doing analyze and forming reaction are of low energy, but of hyper high frequency. They get in to the same field zones of the particles of brain's interface points and form information stream back to the internal matrix. Generally, the front zones, physical fields that surround the particles of the interface points – at level of atoms and molecules, regardless of their functional integrity are able to conduct and transform impulses.

A basic abstraction of the hypothesis is an independence of the codes from their functionality and conducting matter. That means there is no such algorithmic system being dependent of the information which translates processes or memorizes. Each code is a group of code elements which order corresponds to defined bit of information. But that order, respectively, defined bit of information could correspond to optical, acoustic, chemical, organic etc. kind of information, regardless, from or to the matrices.

The memory of an organism contains a large massive of information. Of course, at highest level this is related to the human's memory. Human memorizes unlimited number of scenes, faces, situations, facts etc. Most of them are being memorized and recalled in a motion, namely, dynamical collection and recollection. During its development and training are memorized languages, names, objects with their 3D image and size, people who grow and get aged with their features and many more. All that information is being received, processed and memorized. From memory in an ease or not ease way, instantly or with delay it could be recalled in mind. Considering a pure biological base of integrity of these processes does not seem adequate to their complexity, volume and swiftness of performance.

Up to that point we reached to the concept that each organism consists of a visible part, its biological body, and one invisible, elusive, the integrated to it physical field of third type. The physical field wraps the biological body, besides, the internal organs may have own mini fields substructures integrated to the main field. The points where the field is connected to the material body called gates, bet there are and additional points of contacts, such as the ones of acupuncture procedures. The effect of that procedure is essential, its meaning is to clean and open those points, to improve the contact and the bonding of the field to the biological body.

The field is flexible; it is able to be deformed without destruction or breaking the communications. It is able to interfere with other fields without influence, but in some cases should be affected. Those cases are related to stress situations, energy exchange, and structural deformations.

To maintain the body, the field must have defined functionality. We are unable to say concrete about that functionality and this is not a problem before the hypothesis. Within the traditional knowledge about aura and subtle body there are concepts about its internal structures but here would not comment, following our logical chain. To be functional, the field has to possess the next main divisions, as follows: consciousness, sub consciousness which are abstract but we define their virtual areas, operating of unconscious actions, operating of the organism's functionality, communications, connections to the body and memory. As we said, the consciousness is the outer operating division towards the material world, active, alert during the period when the field is active. It is vulnerable, exposed to influences which could affect and destruct.

The sub consciousness is an inner division, closed to the memory. It is protected because of its functional position.

The sub consciousness operates with records of the standard structures and during the process of relaxation of the field performs recovery of damaged ones. It analyzes the structural status, foremost of the consciousness and signals when diversions occur. But, towards the sub consciousness negative information could be directed which to cause damage of its structures. In cases like these the strong, health sub consciousness forms signals to eliminate the destructions. When it is not strong and health and the damages are in deepness, it is unable to repair. To restore the structures of the sub consciousness is more difficult than these of the consciousness.

The functionality of the field includes divisions operating the processes and the unconscious actions. These processes are biological and exist by all the organisms at different levels. These processes are a subject of analyze the information from both the matrices and foremost the internal followed by forming of operative impulses as a feedback. The unconscious actions are typical for the animals where could form a surprisingly complex behaviors.

It allows the animals to be incorporated in the environment, to create societies at some level in to herds, flights, clusters, families. Some of them, possessing too complex algorithmic systems, perfect sensors, produce complex biological products such as honey, cobweb, toxins etc. To do that, complex algorithms are integrated within their field's zones of unconsciousness. They are obeyed to perform what algorithms command them. Therefore, all complex activities they do and repeat involuntarily are being motivated by the same integrated codes which command and their internal matrix, mean organic processes. Being coded, they do not need of mind to do that. Simply, their elusive algorithmic systems accounting the environment, time, weather, climatic and seismic changes, earth's gravity and magnetic fields and others form command impulses to their executive organs and adequate actions are being performed.

By humans such functions are being saved, foremost in their reproductive system. The code system of controlling field knows almost everything about the status of a biological organism, but the mind and knowledge of humans are an additional factor for its control. Reason actions when standard codes are unable to solve the problems.

The building of all biological structures is a reflection of the functionality of the field. That is so to say: all such structures are coded in the physical field and exist in to it, besides the field incorporates and develops them as biological structures of the organism.

In the context of our abstract considering we could set the functionality in to an order of classification or hierarchy. In the base of that hierarchy are the simplest physical and biochemical structures and processes, the mid floors take the complex ones, and at the top would classify senses, feelings, emotions. It is problem to believe, the matter to be able to form all that, especially the emotions. It is out of doubt, the matter forms the sensors, but the complex code system is a priority of the physical field of third type with its hyper high parameters, foremost, tact frequency, large memory. People ever have talked about soul and spirit and now could give a rational explanation about that term, idea. In that way, the spirit takes realistic outlines instead of the mystique that haunts in the human's considering.

We said, by a growing old human organism only the reproductive bodies – the ovum and the spermatozoids stay not affected by aging. To this would add the changes of the biological stuff of each organism by the processes of metabolism. In each moment a biological body consists of different material particles because in the previous instant has voided a portion of them. But, the personality, characteristics are being saved. Where is here the matter? More, with the process of growing old, at any age, the wish to live is ever preserved, excluding psychological diversions or diseases. What supports the personal identity not changed and alive is the code system of the physical field which remains unchanged. For the standard formula and its tolerances the earth's life of each biological organism is negligible.

The physical field integrated to an organism forms it and its personal identity and exterior characteristics. Diversions of these parameters occur by structural deformations and damages which in some cases could be irrevocable, by damages of gates and connections between the field and the organism. Dependently on the character of the field, it forms the character of the human and his exterior characteristics. The characteristics of a criminal could be distinguished from the ones of a spiritual type in most of the cases.

Have you seen old pictures, from the times when old people had been young and nice? Old time movie, when your favored actor and actress were world sex symbols?

What happened with these people? They are not the same as they were at the picture and the movie? Is something saved from these times? Yes, they are not the same but there is something saved – they are saved! Why they are saved when they are looking and are growing old?

Yes, all people save themselves regardless of the ageing of their material structures. The factor representing a person from the moment of fertilization up to the very end of life and remaining unchanged is the field.

PARANORMALISM

Considering profoundly, there is no rational explanation of life's secrets – how life is originated, developed, and how exists. Most of the people are religious and believe, we are created, but do not know how, others are skeptic and believe only in matter. Because of this, people aim to give to the unexplained phenomenon either, scientific or supernatural expounding. The hypothesis claims, nothing could be supernatural, but everything, every phenomenon and everyone is obeyed to the physical laws and regulations. Here, the problem does not regard the idea of which initiative and creation, but of its concrete realization. The task of the hypothesis is to throw light on these problems giving some universal definition of what is paranormal.

Yes, paranormal within the context of our concepts exists and there are a plenty of examples. The hypothesis defines to be paranormal some diversions from standard algorithms, breaking the gates of communication between elusive and biological bodies and information streams out of their nominal directions. The controlling fields of the people are strictly identical within their Standard formula in accordance with the Creation. They have differences which are principle deflections from the standard within the tolerance formula. Most of these deflections are being foreseen and are in the framework of the tolerance. They are result of the large number of combinations available by the origin of an individual and the various conditions of life.

PATHS OF THE INFORMATION

The path of information is defined by strong laws and regulations applied in to the standard formula. To proceed in the way to distinguish normal from paranormal, we need to do a detailed review of the information streams.

We have defined external and internal matrices of information. The information from both the matrices, external, acquired by the sensors of organism and internal, obtained from its biological structures, via the nervous system, is being sent to the brain. According the hypothesis, the information streams do not stop here, but continue to the corresponded departments of the controlling field. The back information is divided into two basic streams – via consciousness and command impulses sent directly to biological organism. The consciousness is able to command actions of the organism such as motion, speech, to turn sensors to defined directions and many more. The stream, directed towards the organism commands all internal biological processes and involuntary actions.

Those paths of information could be considered to be standard and obeyed to strictly defined regulations. The gates between both the media does not allow information to flow from material world through controlling field to Space Information Field and backwards. There is not allowed consciousness to command controlling field, but only the biological.

All this is valid for each normal organism. When the streams of information sake of some reason have changed their paths, paranormal events occur. The paths of information are being disrupted when some of the gates are being destroyed.

DEFLECTIONS OF NORMALITY

Two basic deflections could be distinguished. The first is related to breaking some of the gates and gap between the Medias. The second is related to diversion of the paths of information.

Each of us is designed to live and to be socialized in the environment of the material world, knowing nothing about the parallel world. Breaking the gates means contact between the consciousness and the parallel world of hyper high frequency.

The living creatures are physical fields of third type. Some of them live independently; others are integrated to material organisms. The physical fields of third type algorithmic systems include functional independent fields which billions years before had implementing the Project and fulfilled the Creation on the earth. We may imagine everything round of us pierced and permeated by hyper fields of third type of different kind, including the fields integrated to the plants of the jungles, forests round the planet and these integrated to the numerous animals inhabiting these large areas.

In some cases some of these creatures are closed to us living in symbiosis with the controlling field of a given biological organism.

Such independent creature never has had material, biological or other organism being attached to. Its code system is organized at very high level of intelligence. Has no linguistic problems or barriers to communicate to the organism to which is in a symbiosis.

These creatures have the advantages that give him the Media – hyper quick communications, penetrating through the matter of the material world. It has no problems to bring information from the opposite part of the earth for an instant. It has communication and access to the Space Information Conducting Media and to its hyper large database.

As far as from the Center of Space Information Field define our destiny and future, or there is large memory related to our past, these creatures are informed about.

When the gates between the controlling field of someone are somehow broken, information between the Space Information Field and the consciousness of the individual becomes available. This is one form of paranormal. The Author and his family were engaged in such a case from 1991 to 1994. Result of the communication between a woman and non-material creature called Mo are plenty of paintings on canvas, made by fingers. The woman has had never before those skills and never been an artist.

The non-material creatures take place in such occupation as a rule after coma, when the physical field, integrated to the biological body leaves the organism during the time of coma. The creature perfectly occupies some of the gates, and after return of the basic field, i.e. when the organism comes out of coma remains in closed relations with the individual. The things are more complex and hard to be explained but Mo saves the life of the woman. This is the reason for the unusual behavior of people in coma – living and not living; making motions or speaking even but remaining in that state of unconsciousness.

Probably there is other a form when the gates between are somehow opened or broken. In that case direct contact to Space Information Field, even to its Center is possible. These people have opportunity to communicate to the controlling fields of others, to receive information from unavailable directions. The consciousness of these people is not perfectly isolated from the parallel world.

Most of these cases are a consequence of post – coma or post clinical death condition. In both the cases, the controlling field leaves the biological, but returns back by some reason. On its return, the gates are not perfectly coupled; some gates stay open, uncovered. In that way the hypothesis classifies as first paranormal phenomenon contact through the gates between the material and the parallel worlds. There are numerous known phenomena like Edgar Kasey readings and many more which we do not have a task to comment here.

The second basic paranormal is deflection of the normal paths of information. In that case information stream between consciousness and controlling field is becoming probable. People who have that form of paranormal are able to command at some level their field. Normally, there is no communication between the consciousness and the controlling field. What this means? Means, in a conscious way we are unable to control our field but only the material organism; this is a subject of internal organization.

The abstract part of the field consciousness is not provided with information related to its controlling field. No one knows about its presence, so far that it should be problem to give credence to the written here. From other side that means, the individual is unable to influence its field.

That means as well, the consciousness, as being a code body integrated to the field is obeyed to its algorithms and there is no way to be other. But, in some unusual cases these algorithmic restrictions are reduced, the information streams deflected and some communications between consciousness and controlling field possible. In such cases the consciousness could influence the field and these people give evidence of amazingly abilities.

Such is the man, who breaks handles of tea spoons. When the existence of controlling field is being unknown or refused, the skills of that man are impossible to be explained. That and occurs, the science is puzzled, the man becomes world renown. But, to try to expound that in the context of the said above and the picture should be as follows.

The Hypothesis does not give detailed description how the controlling field is constructed, but, the fingertips are antennas and important information and energy channels pass through them as communication between the controlling and biological bodies. These communications are better by biologically healthy fingertips. By destruction of their structures, for example, by mycosis, they are getting make worse communications.

The man who breaks the handles of tea spoons has not only excellent communications via his fingertips, but he is able to command his field by consciousness. He deforms his field, concentrating stream of energy between fingertips. As we mentioned no once, that energy is of hyper high frequency, unavailable for our material world. The hyper high frequency of the field applied being instantly concentrated at a thin section of metal puts its particles in to a hyper swift motion. As result, in that section the forces of adhesion between the particles are broken.

In that case as well by all paranormal the controlling loses energy. Especially, when the consciousness commands and deforms the field, the latter not only loses energy, but it breaks its normal functionality during the experiments. After a number of experiments such field is unavailable to rehabilitate its structures and energy and needs long relaxing period by special conditions. Of course the energy needed to recharge has nothing common with the energy of the material world and the source of that energy is some special place.

INFLUENCES ON THE CONTROLLING FIELD

As it was expounded, each controlling field is provided with function “Protection” by means of which its integrity is preserved and influences and information transfer between the Medias – prevented. On controlling field influences from both the Medias are probable.

Influences from the material world could be detected and the field to produce towards them emotional or / and actual reaction. These influences are registered directly and reaction is conscious or subconscious character. Such influences are permanent during life and result of effects of material or social environment on organisms’ sensors, or by interference with controlling fields of other people, animals and plants. These influences could result on the structure or energy status of any controlling field.

In some cases destructive effect should occur, for example, by stress. These influences are temporary and during a process of relaxation the field could restore structures and compensate an energy loss. But, by long lasting active term, respectively, shortened the time to relax, the compensation is a problem and the field alerts the consciousness. In some cases these influences could have positive effects of relaxing - restore structures and improvement of the energy status. For biological body that means improvement of status of health. These influences are probable by contact to some animals and plants, generally, to creatures which possess strong and healthy field structures with excess of energy.

To finish this topic would mention the only two factors which could affect directly the physical fields from a side of the material world. These factors are thought and ionizing radiation.

The thought is an emanated from the physical field portion of energy that contents coded information. While in our material world we communicate by means of language, in the parallel world the language is the code system. If one who emits a thought has a high energy status that is very rarely and is considered as being a phenomenon his thought could charge objects with energy and information and to affect other physical fields. As well there could be created information bridges between fields or fields and objects.

The ionizing radiation actions on the tissue, but and on the physical field causing deep structural damages. As a consequence the field becomes deflected from the standard, and loses its normal functionalities. It becomes unable to build and keep the biological body according the standard and makes that out of it. To this are owed mutations that are namely a material expression of that nonstandard origin and development. Here we do not have in mind the effect of the ionizing radiation directly on the tissue, causing its simple destruction.

All other electromagnetic fields produced by the particles of inert matter have parameters far from these of the fields of third type. Therefore, they are unable to affect directly the fields integrated to organisms, but probably the zones of interface or their effect is not essential.

INFLUENCES FROM THE PARALLEL WORLD

Quite different is the influence from the side of ethereal media. Such influence means direct contact either, with Space Information and Energy Field, or with living creatures not integrated to material organisms. Influence from the Space Information and Energy Field could be long lasting and to cause a qualitative change the status of controlling fields or receiving some unusual information.

Exodus 31

Bezalel and Oholiab

1 Then the LORD said to Moses, 2 "See, I have
chosen Bezalel son of Uri, the son of Hur, of the tribe
of Judah, 3 and I have filled him with the Spirit of
God, with skill, ability and knowledge in all kinds of
crafts- 4 to make artistic designs for work in gold,
silver and bronze, 5 to cut and set stones, to work in
wood, and to engage in all kinds of craftsmanship. 6
Moreover, I have appointed Oholiab son of
Ahisamach, of the tribe of Dan, to help him. Also I

> have given skill to all the craftsmen to make
> everything I have commanded you..........

This is one of the plenty cites from Bible, where Spirits Santos influences on somebody's controlling field to achieve skills, knowledge and wisdom. The relation between Hypothesis and Religion is a subject of a theme that would be reviewed further in that part and in part three, "The Evolution".

History shows plenty of cases when distinguished persons demonstrate qualities, skills and knowledge that are not feature of their époque. From the aspect of their époque, what they can and know is amazing. Of course, the most distinguished is great Leonardo. His works are well known, for the hypothesis in the context of the current topic life and work of the geniuses is of particular interest.

It is known, his projects have nothing with the level of technology and knowledge of his time and most of them are realized centuries after. The Hypothesis claims, he has acquired exclusive influence and information from the Center of Space Information Field. Why – the answer is in the third part. The same exclusive influence gets the great Renaissance geniuses of arts, physics, further in technologies. The great Napoleon is not only a commander, but and a creative individual. Other personals from the past century are distinguished as politicians, dictators influenced the world history. All they are raised from ordinary to highest social positions being distinguished from the others. Not all of them have had these qualities by birth. These qualities have acquired in a moment of their individual development, besides, the influence on them is ever purposefully. They are selected not by chance, but with a precision that to guarantee they would receive the influence and realize their purposes.

In the context of the hypothesis where these processes are being reviewed within an impartial manner and pure physical plane, follows to expound these phenomena. It is influence on their controlling fields, adding codes, but with one small detail. Influence on personal field is out of regulations and possible only by some breach of their standard code structures. In some cases this should lead to some level of deflection of individuality. That deflection could be not only consequence, but and prerequisite to achieve intervention. Such are psychology, sexual, healthy, diversions from standard behavior. A prerequisite should be as well long lasting exhausting of organism, respectively, insufficient energy and breach of structures of controlling field.

All these prerequisites or consequents lead to the same result – opportunity in the controlling field of individual a gap to be made through its protective shield and the information to be delivered. That influence is related as well and to energy factors. During its performance the energy status of the field is being weakened. But, after each intervention that status immediately has been restored and improved by energy charge.

And in addition, about the heredity of the geniuses: are geniuses generations of the geniuses? There are not or extremely rarely.

MAGIC WHETHER IT EXISTS

Magic and sorcery are the influencing of events, objects, people and physical phenomena by mystical, paranormal or supernatural means. The terms can also refer to the practices employed by a person to wield this influence, and to beliefs that explain various events and phenomena in such terms.

One of the most widespread magical procedures for healing, harming or otherwise influencing someone from a distance involves making an effigy of him or her from any material. Actions performed on the effigy are believed to result in analogous effects upon the target person, so that, for example, a part of the effigy's body may be damaged in order to cause pain or disease in the same part of the target's body. This magical technique may be employed for maleficent or beneficent ends, and even for giving help to gods against malignant demons.

Another similar procedure by which an enemy can be injured is to gain possession of some of the enemy's hair, nails or other bodily by-products, or a piece of clothing, and treat these in some hostile way.

Although these magical actions may often prove effective, this can in many cases be explained by placebo, without requiring a paranormal explanation.

External influence should be reviewed and in the context of existed traditionally and even in our modern society superstitions like believe in black and white magic. It seems people could by speech or thinking to activate unknown forces from the Parallel world to intervene in somebody's field and to change his life.

Example in that context is the benediction. It is ever related to our Creator to make good to people. A lot of examples could be found in the Bible.

People with malice or egoistic intention try to influence the destiny of others. People who are religious rarely should be influenced in that way, they have a strong shield.

What people call magic is an influence on their controlling field by which structural changes become. The hypothesis distinguishes three basic code areas of control integrated in to controlling field. Namely, consciousness, sub consciousness and control of organism's biological processes. Consciousness, being outer and actual area is less protected against influences. It is divided in to sectors distinguished by their code sub integrity, respectively, functionality. Each sector is related to defined area of activity, such as body and organism, individual's social relations, private and family, professional etc. The sectors are relatively independent one to other. The external influence is directed to defined sector, where commence changes as follows. First is going destruction and the sector is no more able to action as before. Follows basic substitution of that what has been destroyed. After the basic change comes implementation of new algorithms to replace the sector. The last function is to protect the new sector against return to previous, normal status. As weaker is the shield of the controlling field, so stronger is the negative protection. Further the influenced sector actions simultaneously with the others, unaffected ones as result of the said relatively independence between them. The individual should stay perfect professional but to fall in to family or social conflict for example.

The sub consciousness being internal and closed to the memory area is well protected and could not to be influenced in that way. Being influenced, the controlling field begins to action to return to normal status and the role of sub consciousness is becoming crucial. It sends signals to consciousness attempting to activate it. The return is difficult particularly by strong shield of intervention. Sometimes the intervention ant its shield is not strong, but the return also, in that case follows long lasting period of wavering.

Most of these influences are tangible as signals from material physical cause, but part of them is hidden for the consciousness and to be revealed must be explored. Controlling field is provided with algorithms for diagnostic of influences from the material environment, but, there are changed conditions that are not being foreseen. Such are increased radiation, electromagnetic fields, modified foods, artificial materials. In these cases the code system is unable to perform reaction and resistance. These influences are objecting of exploration by human's reason their influence to be reduced or suppressed. Unfortunately, the same reason creates new sources of influences.

An important physical factor is the energy status of controlling field that determines its ability to make diagnostic and to create effective reaction of resistance. A result when the interference requires more energy for its suppression than available is energy misbalance, reduced functionality, neglected activities.

The process of aging was a subject of explanation, as being a permanent consecution of influences that requires more energy. Insufficient energy gradually leads to become rarefied the information streams and neglecting of biological processes. During the time these processes become steadily and the energy needs increases, respectively, its insufficiency. The field which performs complex activities to process both matrices falls in difficulties more and more. It must keep organs in their 3D positions, in the process of metabolism to restore the tissues as the original requires. By reduced density of information that space positioning is unavailable to be in control precisely and in biological body particular changes occur. Such is for example dislocating of eye balls from their nominal positions and the consequences of that. Change becomes in all organs and tissues that are well known.

HYPOTHESIS, RELIGION AND PHILOSOPHY

Religion is always being related to something strange, the belief. Why must believe or not to believe? In what people believe and why some of them do not?

Believe regards something undetectable. Part of individuals say it exists, other part says, it does not exist. In that way, who really should believe in that hypothesis where everything is invisible and rejected by the official concepts? Let imagine, there is a scientist skeptic, who is believer, surely, there is one. He or she believes in what?

Considering religion as being a form of manifestation of human's awe to supernatural forces that have created us and owing to which we exist, the hypothesis gives synonymous to the question to believe or not to believe. The Hypothesis regards with respect to all religions and their rituals but approaches to that topic purely pragmatic, based on the physical laws and regulations. Despite, we consider, for our Creator nothing is unavailable. After all, and the hypothesis, despite of its daring statements is limited by a human mentality. But, it gives a synonymous answer – Yes, we have been created.

Further in part three "The Evolution" is asked and the next question why, and it would be considered in the framework of the author's resources. That question is closed to other one – how it is possible, humans to origin from apes during an evolution process, and at the same time, to be created. Not only we have been created, but the whole evolution process has been carefully considered by the Great Supreme Reason, and within that chain the human is the crown. In that context, the Hypothesis is interesting about the physical parameters of possibility to act on the matter of our world. As it was stated, the Idea following the way from the Center of Space Information Field through Universal Conducting Media reaches to each point of the Universe. The Space is filled with information streams and integrated algorithmic objects.

During the times of existence, humans have compared Our Creator to himself and ever have searched in Him features of a human. Next question that ever had been actual is why He is invisible and does not appear? That is base of belief and skepticism that go by humans hand in hand. From there descends hypocrisy with which some of people relate to believe.

From point of view of the physics a transition between ethereal and material Medias is too complex. Both Medias have principle distinguished physical parameters like frequency which cause problems by energy and information transfer. The hypothesis is unable to interpret or to claim to give comprehensively all answers in that location in deepness. In the first pages of that material a framework and opportunities of the hypothesis were defined, out of which we do not like to explain what are unable to do. What is of interest for people about religion is excellently developed and narrated in the Bible, with great respect to which, as well with a level of knowledge was written the current material.

The philosophic categories which explore the human existence give some explanation of its meaning. Within the hypothesis a considering was developed about human creation and structure which should seem slightly to degrade his creature and even contradicts to contemporary comprehension about. The Author sincerely hopes not to be thrown and burned at the stake in 21 century.

But, what was expounded should be accepted as being logical, or even, the only possible, in which of course, the author is deeply convinced. Does such a formulation lower the prestige of human as being the only reasoning creature? Not, at least. It was in details expounded: communication between both the Medias is prohibited. The consciousness of human is directed entirely to our material world in which he lives and creates. His controlling field and algorithms integrated in it had been designed in the way making human absolutely independent in thoughts and actions. There is no way to compare human to a robot leaded by some external force.

By the process of reproduction the generated controlling field is not exact copy or simple compilation of parent's, as it is by plants, but a selection of their features. It saves more of parent's features, but during the new life is an object of development, cultivation and education. The new born human controlling field needs a lot of information like memorized experience, skills and knowledge as an indication for independence of human as being a creator in his material world. What are then his restrictions and destiny, simply, open the Bible and read.

HYPOTHESIS, SCIENCE, FANTASTIC

Evidently, the Hypothesis would fall into conflict with contemporary official scientific and public considering. But, the problems, subject of its essence are still avoided to be realistic and rational expounded. It should be expected the material to be qualified as being esoteric or speculative. We hope the train of reasoning and logically substantiate to be in behalf of Hypothesis.

The hypothesis explains human consciousness directed to the material world. In that way we are unable to become anything from and about the parallel world and this is prohibited for us. More, what we know about other a dimension of much higher frequency of vibrations is nothing or ancient knowledge or legends. This motivates the human reason to deny any possibility anything outside of what is visible and sensible to exist. But if we are unable to explore a phenomenon does it exist or not? Is the phenomenon obeyed to our concepts and terminology? How life exists without our concepts from billions years? Is the concept about objects to attract one to another right? Is the concept considering the elusive media to press the objects one to another less probable? But, ether is undetectable and therefore it does not exist, the space is empty or filled by some abstract considering.

Panteley Bahchevanov

BIOLOGICAL MODIFICATION AND SYNTHESIS

One of the problems before contemporary science is that of genetic engineering and modification. Crown of that aspiration is the aim a biological organism to be created. What is genetic modification? This is an influence in location where preliminarily standard codes are being set.

Science development at atomic and molecular levels allows action closed to the micro world, biological micro bodies to be well explored and intervention in genetics. The genetic changes are based on considering the material world is the only available and the matter is self-organized in to biological structures. According the hypothesis, each genome and its information are structured under the codes of controlling field and for a given organism have defined functionality. In that way, the organism acquires features that define its behavior and position in the chain of species and ecology system.

After intervention these features should be changed in some undetermined way and supposing, its behavior becomes unpredicted. Surely, a conflict between standard formula, coded in its field and modified biological structure should be expected. Effects of such actions are stable and fast grow animals and plants, but without flavor, non-aromatic flowers, unknown viruses. We do not evaluate that activity and deeply believe the scientists are mostly reasonable and cautious to prevent any negative results of it.

It is quite different the situation by attempts living organism to be synthesized in laboratory. To achieve this, effective metabolic system must be created and all to be set in a biological package. There are some problems, foremost, the operation of processes, even of the simplest microorganism are too complex. Unfortunately, at that point we must overcome the dogma that matter is self-processing and to think seriously for some system to operate it.

There are two ways to do that, to arrange a material or elusive operating system. In the first case, to the artificial virus should be attached a micro processing unit and it would be looking well bringing that device on back. The second case requires the ethereal media to be recognized, explored and us to be able to create integral code bodies for it. Currently this is a fantasy. Seems, to create simplest biological organism is not the same as to dig and modify genetic systems, the barrier of physical parameters of both media is too serious.

CHIP IN A HEAD

Plenty of fantastic stories and movies consider the topic of modification human reason and consciousness by means of a chip implantation. Usually in such stories aliens capture representatives of human race, put under high tech operation effect of which is implantation of a chip in brain, acquirement of new consciousness. We keep in mind, our material media is restricted by speed of light. To travel in space by highest speed is impossible for each material substance, even made by high tech advantaged civilization. To aliens would take a lot of time to travel to our planet, but let suppose that is being realized and let return to chip.

It is a material integration which is being coded, no problem to be energized. Its operating impulses must be directed to the gates of brain precisely and to correspond to the information stream that would like to affect. There should appear inconformity between the current controlling code system and the material integration. All that does not mean, a chip would not affect the brine at all, it should be expected an intervention as disturbance to occur. The result expected could be an interference with the basic information streams and operating impulses.

At that point would like to open a parenthesis. Considering of people is mostly related to material media that is well known, besides, we imagine and would like the unknown to be as the known. The aliens are obeyed to be material creatures, but we neglect restricted opportunities to travel to long distances in Universe. It is strange our wish they to travel by flying saucer. The latter absent in our technology arsenal, because are aerodynamically unstable, but are very suitable as obligatory attribute of aliens.

Of course, the worm hole is coming to make traveling at hyper long distances through the Universe ease and short. According this concept the clusters at billions light - years might be closed to us. But the hypothesis does not comment this topic.

COMPUTERS, ROBOTS, HYPOTHESIS

It becomes apparently an analogy between computer, robot, CNC operating device etc., and the claims of the current material. One should say the hypothesis has used that analogy, implementing the bases of computer design and operating of various devices, such as industrial controller, robot and CNC machines by impulses from microprocessor. Yes, it is same; there are algorithms, integrated code bodies, sensors, streams of information and executive devices. The logic is the same and it is worthy to follow that chain. It would be a topic again in the context of evolution in part three.

It is obviously, the simplest organism created billion years before has principle same system, as an organism today. Essential difference between a computer system and that from the claims of hypothesis is the integration – on material, hard base and within the environment of ethereal Media. This distinguishes both systems by physical parameters and abilities to be fast, to process and memorize large and hyper large quantities information. At that moment the artificial systems are far from the ethereal – biological ones, they are limited by the physical parameters, available for our material world. Other a question should arise in that context, is human a robot? That was expounded above, human is creative and far from this to be a robot.

THE GREAT BEYOND

The Hypothesis defines a parallel to our world, in to which we are immersed. In that context follows to try to explain the relations between both the worlds. It would be done by an abstraction but and analyzing some facts in the framework of a logical and admissible possibility.

It is known, some of the famous people of our present time have had as consequence of healthy problems experience between both Medias. They have changed them after leaving and consequent return in the material world. The science attempted to convince them, all have been illusion, produced by their biological protecting systems in the critical moments. All of these people have changed their thinking and philosophy and nothing could cause to hesitate.

The author has discussed topics of that material with Ms. P.S., distinguished physician neurologist and psychologist from Varna, Bulgaria. It turned out that she, being under anesthesia has had a critical moment.

Cite: "I went in to a cylindrical tunnel with clearly distinguished spiral thread by its periphery and was put in accelerated motion in to it. The return was suddenly."

She has been leading away of that dangerous condition by the medical equip besides her organism overcomes the crisis. In the essence of the things, the controlling field of Ms. P. S. has evaluate the situation, mean, the information from the internal information matrix and decided, it is unable to overcome the problem with biological organism. Closely in time the medical equip provides successful activities to return the organism to life. Information from the matrix starts again and controlling field, being not inertial returns rapidly to the organism. In that case all gates are being restored properly and no negative effect for the woman occurs. Seems, her controlling field has had the wish to make a little journey in the space.

Let return to the misty beyond and continue our explanation. It is clear; our biological life on earth is limited. The controlling field needs relaxation to restore its affected by active life structures, as well, its energy. That is performed normally by sleep, when all sensors are inactive, the communications between the field and organism interrupted, leaving active only ones of existence importance. Besides, the field, being flexible and non-inert has opportunity partially to be separated from the biological.

Then its communications to the ethereal media as well internal relaxing processes are being more sensible in a form of dreams. Similar are processes by hypnosis and influence of narcotics. In these cases the communications are being affected, besides gates that normally are closed are becoming opened. Such are communications to the memory and closed to sub consciousness, from where information is extracted being inaccessible by normal conditions. It is essential the separation of the field by conditions of narcosis, similarly as by the case narrated of anesthesia. It is well known the influence of narcotics by over dosing when the field separates so far that its return is a problem.

The relations between the field and the organism are related to aging, diseases, injures, traumas, external influences. To command the organism, the field analyzes its status at atomic, molecular, cell and macro levels. It performs comparative analyze of chemical compounds and elements that come in and the need of them and signals by in or over sufficiency or simply regulates their content. This is extremely precision and complex action and large information processing.

Biological organism exists by metabolism, besides each particle which has been changed must take exactly the same position by the same conditions and to perform its functions by absolutely same way. When by some reason the field neglects some of these processes it leads to aging after collecting defects in organism. Except aging, the organism falls upon influence of many factors that change its status and that the field is unable to avoid, for example, poison, fracture etc.

After comparative analyze the field defines whether the injury of the biological organism is fatal and no way to be overcome, or there is insufficient energy, especially by long lasting exhausting disease, and takes decision to separate from the biological. By separation it breaks integrity between both bodies and suspends all control and processing functions. Because the processes in the ethereal media are non-inert, the distances not of importance, in some very few cases the field may return, if by its separation receives signal, the problems of organism are being surmounted.

Logically, may take conclusion, the field stays in some level of nonfunctional touch with organism for some time. Probably, a slight thin string between them remains forever. Controlling field, being defined and independent begins own life in the parallel world that is enigmatic for us.

ABOUT CRYOGENICS

Cryogenics is the branch of the physics related to the production of very low temperatures and the behavior of the materials and organisms in those temperatures. Cryonics is emerging medical technology of preserving humans with the intention of their future revival.

The current technique of full-body preservation with cryogenic protective chemicals causes extensive molecular damage to the body. To successfully bring a patient back to life, cryonics would not only need to reverse this damage, but would also have to cure the original illness the patient died from. Science has already discovered ways to suspend and revive biological life forms. Today, relatively simple living structures such as red blood, stem cells, sperm and embryos are routinely preserved using cryobiology technology.

How this delicate problem appears in our context. A biological organism could live only when a controlling field is integrated to it because the controlling field animates the mass of material particles. During the time of integration the field permanently analyzes the status of the organism. When this analyze indicates for heavy, fatal injure as a consequence of decease, aging or mechanical damage it takes solution to leave the material body. That solution concerns the field as being no more able to operate the organism sake of irreversible processes or lack of energy to overcome the problems. The field breaks its communications and connections to the organism and separates of it. The chance of the rest biological mass is only to be decomposed or to be processed by cryonics.

A connection between both the bodies keeps especially at the first few days after separation. In some cases after coma or clinical death the controlling field returns. It is provided in the code system after a positive analyze of the organism's status immediately to return. There is one more powerful factor which decides the separation or return to the organism.

This is the will of God. For us used to perceive God as some abstract or mystical notion this may sound not realistic. But there are cases that author does not like to explain in details when after long lasting coma, the first case and heavy injure after a gun shot, the second, by will of God these people returned saving clear impression and remembering what happened beyond. The second case was from the view point of the medicine absolutely insoluble and revive incredible.

Retaining relation to the non-living organism, its field saves the opportunity to return if conditions about occur. The field exists in the ethereal Media where motion of the particles, respectively, time does not exist and the space is different than this of our notion. There are conditions the field that belongs to the organism to return and after 50 years. But, the cellular structures of the biological body must be preserved without additional damages caused by the cryonics process and to be cured from its previous illness, injure or deep aging. The question is about how far the organism could be driven to optimal condition and to afford opportunity to the field to operate it effectively. Finally everything would be decided by the will of God.

CONCEPTS REGARDING THE CONTROLLING FIELD

There are traditional comprehensions about aura and chakras. With certainty they reveal the structures of them, areas, layers, functional points. The hypothesis is constructed, following defined principles and avoids entering in locations, where the abstract and deductive considering are being replaced by unknown concepts. In this way, the hypothesis rejects to comment the said comprehensions.

The hypothesis makes a logical assumption, following its train of reasoning. It reveals the controlling field as consisted of integrated codes that define functional areas, basic of which are Memory, Control of organism, Sub Consciousness, Consciousness, Communications. The sub areas are diagnostic, comparative analyze, protecting, information processing.

LIFE SUMMARY

And here we would systematize the material, generalize the most important and draw conclusions. This is and an abstract of the material.

At first we put the base of life on existing of ethereal Media. Further approached the conclusions, the biological structures as extremely complex upgrade of the matter are impossible to be self-created by the material particles.

In that logical train follows the probability and necessity in the ethereal media to exist independent self-integrated code bodies with distinguished functionality which penetrates through the matter. Most of them are not integrated to biological organisms from material world. These, who are integrated to organisms control them and make life possible.

Controlling field (field) controls all processes of the biological organism. It structures and develops micro and macro biological bodies such as cells and DNA, hormones, organs, systems according the algorithms implemented in to it. It creates complex and at high level precise biological structures, such as optical and acoustic, chemical analyzing sensors, analyzers of various environmental physical parameters.

The field is defined by its structure and energy level. In active period it endures destruction and energy exhausting. During the sleep it interrupts the communications between both bodies, leaving only these which support the metabolism and in that way restructures and charges energy. During sleep the field is partially separated from organism.

The hypothesis divided the information stream to the field in to two basic matrices of information. One of them which correspond to the organism called "Internal" and the other "External" which corresponds to the information received by all the sensors from natural and social environment.

Existing of the simplest organism requires processing of a large stream of information and analyze at atomic and molecular levels to support metabolism growing and all other internal processes. Talking about human, the information from both matrices gets in to determined areas of controlling field correspondingly to its nature and location of acquirement. The field makes comparative analyze and performs emotional or functional reaction via consciousness or directly to organism. The back information that controls organism is being transferred by means of hyper high frequency weak and fine impulses to the gates in brain, central and peripheral nervous systems.

Two physical parameters of the field are basic and crucial – energy and tact frequency. In ethereal Media there is a lack of energy; therefore, the field charges properly transformed energy from material world. That energy transformation could be performed by other ethereal bodies, including specialized to do that as well by unknown sources.

The tact frequency determines the opportunity to process hyper large information streams at hyper fast, non-inert level of reactivity and to realize hyper large memory. The physical parameters of ethereal media allow algorithms of highest level to be realized, including emotions and reason.

Each controlling field of third type, integrated to and operated an organism possesses algorithm to copy its features. The average of copying corresponds to the level in the hierarchy of species. The plants copy entirely genetic information and reproduce parent's features as they are. In difference, by the animals and especially humans, copied genetic information is selection of the parent's. Each macro body, responsible for reproduction activity, such as ovum, spermatozoid, and seed has integrated to it controlling field that contents record of genetic information, copied from the parent.

It possesses all algorithms needed to origin new organism and for its future development. The field of each sex cell of bisexual organisms, such as ovum and spermatozoid possess the information from one of the parents, male or female, but it is not enough to origin new life. By the process of fertilization the information, carried by ovum is being combined by that of spermatozoid, creating new controlling field that is able to origin and develop organism. The genetic information content in the biological structures, DNA is reflection of that of the ethereal structures.

It is a phenomenon how from a small ovum and microscopic spermatozoid a living organism is created, able to grow and develop. It seems excluded that to be possible without algorithm, set once forever. By process of growing of an organism and other stages which the algorithmic system knows perfectly how to build in space and in time its micro and macro particles in defined order, timing from the moment of birth to last moment of life. During life environmental information is being added; acquire experience, skills and knowledge, but basic codes stay unchanged.

The lasting of organism's life is defined by two basic factors. They are previously coded duration and aging. During life, each organism experiences unlimited number of influences, including invisible microorganisms, stress, environmental, social. To overcome the consequences, the field needs more energy. By young organism these influences are still not essential number and to be overcome is possible by less energy consumption. During years, number of defects increases in a mathematical progression, the energy to be overcome too. When the field is unable to charge that energy, it begins to rarefy the information streams and command impulses. As consequence, the organs that receive less command impulses become partially neglected. They change size, structure, shape, position, functionality. For example, the eye balls become destructed and moved from their nominal position that corresponds to worsen eyesight. Additional influences, such as heavy diseases, genetic injuries, environmental and social could accelerate that process.

For each organism the duration of life is being limited. For human that barrier is up to 120 years, reference in Bible. Approach that age is prevented by the reasons for aging, described above. By approaching of defined status of organism its controlling field is no more able to control it and disintegration between both bodies becomes. Separating of the field from organism means end of its life. That separation could be temporary, in rare cases, including coma and clinical death. In those cases, the field returns, restoring integrity, but some of the information gates between field and brain should stay opened.

The longest part of "The Parallel World" was dedicated to life, such as we see it and such as we could not imagine. The science is advanced too much in exploration of the biological organisms at atomic and molecular level and undoubtedly would continue in that way. But, there is a barrier of the Media which we still deny because are being unable to detect. The absence of ethereal media with its incredible physical parameters in the scientific accounts leads to mysteriousness of life at all. And only that media with its physical parameters could give us the only explanation available.

ALL ORGANISMS HAVE CONTROLLING FIELDS

All biological organisms are material, but they could not be conceived, developed and to exist without integrated controlling fields that to control them. The plants have root – trunk; supreme animals and human – brain – spinal organization of the fields. The seeds that are able to germinate have fields up to the moment of their damage, when are no more able to. The fields, integrated to ovum and spermatozoids are not complete –

they become after fertilization, creating integrated field of new born human organism.

CONTROLLING FIELD INTEGRATED TO HUMAN

It is represented as grayscale intensity zones, besides the intensity corresponds to the one of field. Round human's biological brain the field is most intensive. The field is flexible, spread far from biological body to which it is integrated.

EXTERNAL MATRIX

External matrix is represented as picture that human perceives visually and audio signals. In addition, external matrix comprises physical information from all receptors and sensors. At the illustration, a young organism perceives full information from all external sources. During aging the information matrix reduces its density, shown by grown pixels image (early stage) and by grown points (late stage).

INTERNAL MATRIX

The information from all internal organs and systems of biological body is being received by controlling field.

Signals that operate organism are being sent back. By young organism the matrix consists of full information stream.

INTERNAL MATRIX - AGING

During the process of aging organic defects are being accumulated and cause problems of controlling field to process information and to operate organism. The latter neglects part of information, respectively, the stream of control signals. That is represented by grown pixels.

INTERNAL MATRIX – ADAVANCED AGING

The information stream and control signals during advanced aging process are represented by grown points. The information through matrix is the same, but most part of it corresponds to defective organs, which problems must be solved by controlling field. The process grows in a progression.

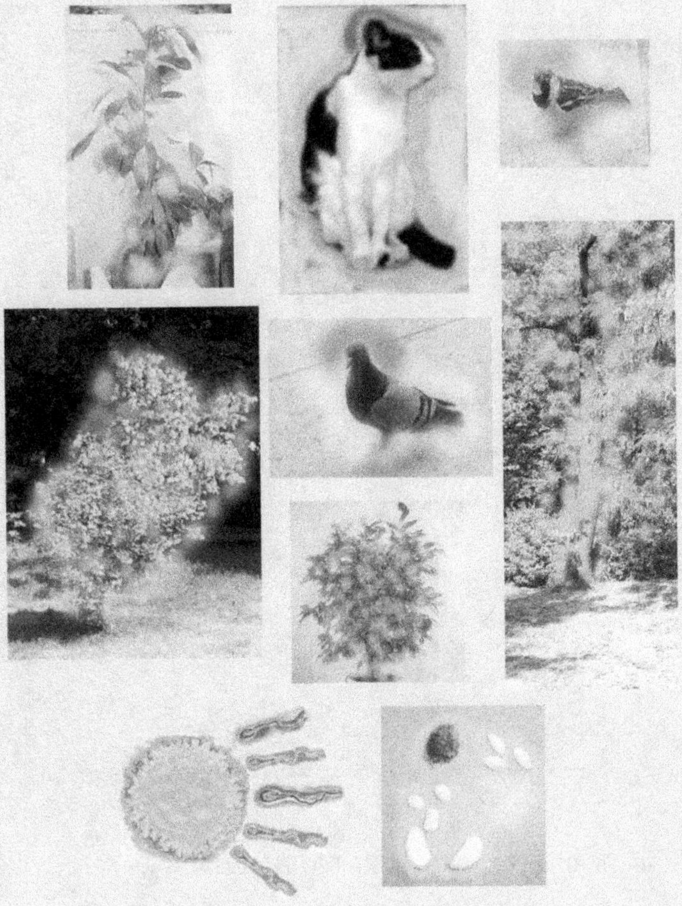

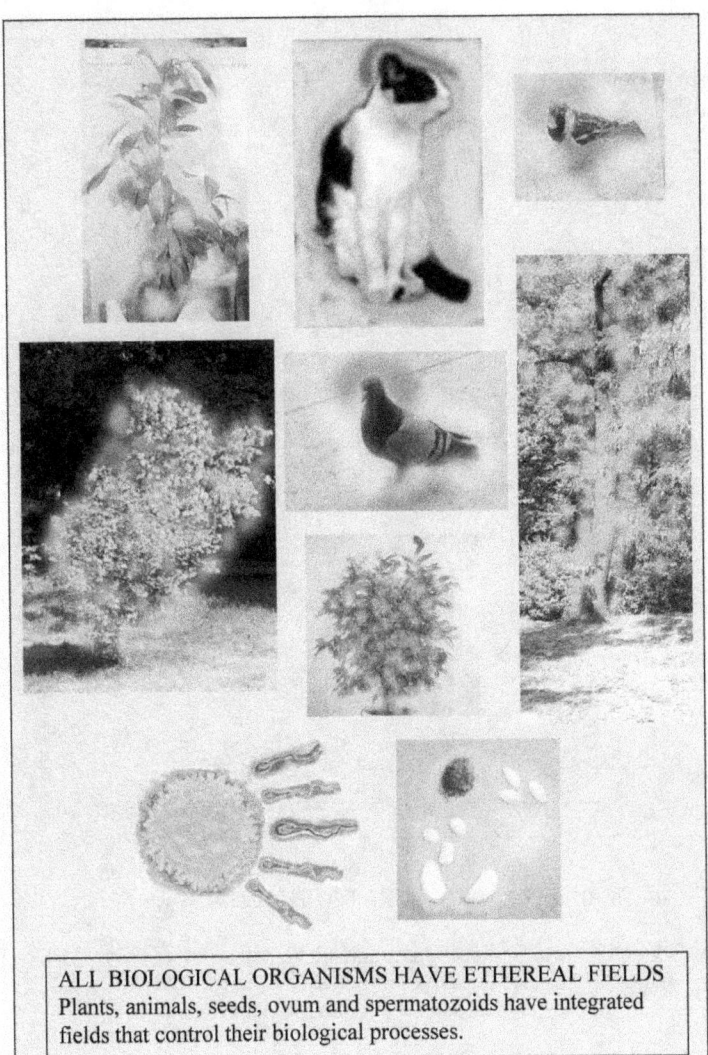

ALL BIOLOGICAL ORGANISMS HAVE ETHEREAL FIELDS
Plants, animals, seeds, ovum and spermatozoids have integrated fields that control their biological processes.

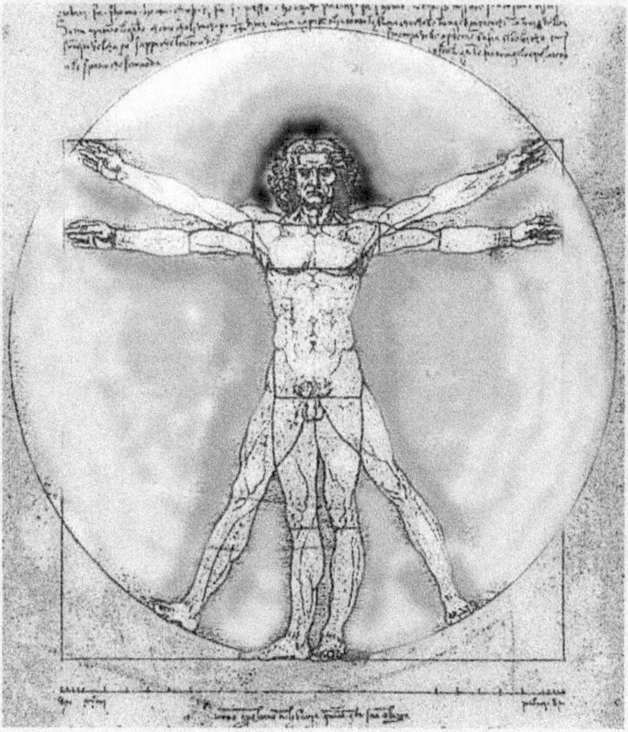

ETHEREAL FIELD INTEGRATED TO HUMAN

The field is most consistent round the biological brain, from where through the gates it receives information from external and internal matrices and sends controlling signals back to it.

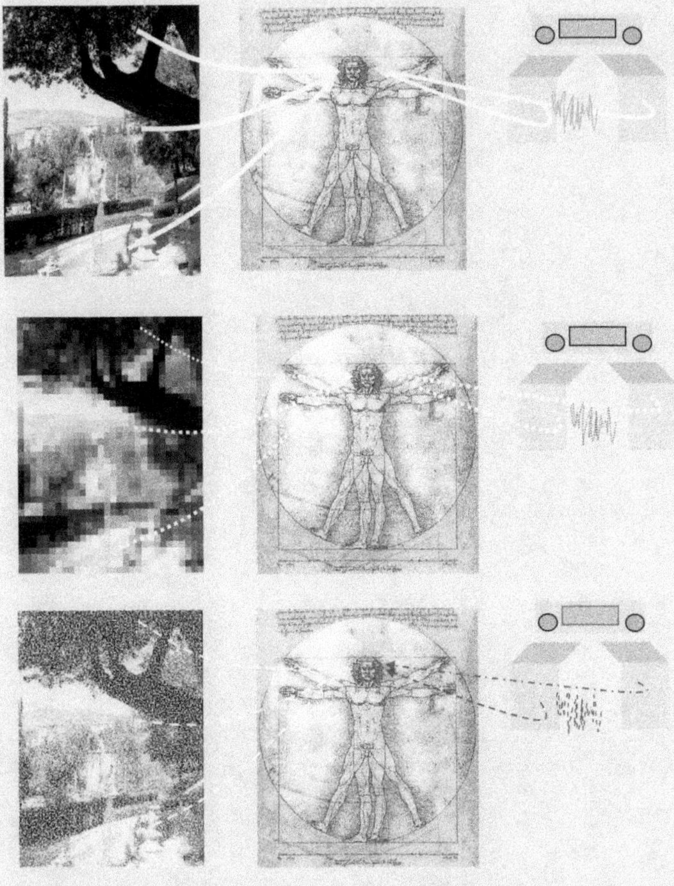

EXTERNAL MATRIX from visual and audio signals
During the aging the information stream from external matrix decreases density. In addition to visual and audio are all sensors that transform physical environmental information. It is being translated via peripheral and central nerve systems to brain, via gates to the field.

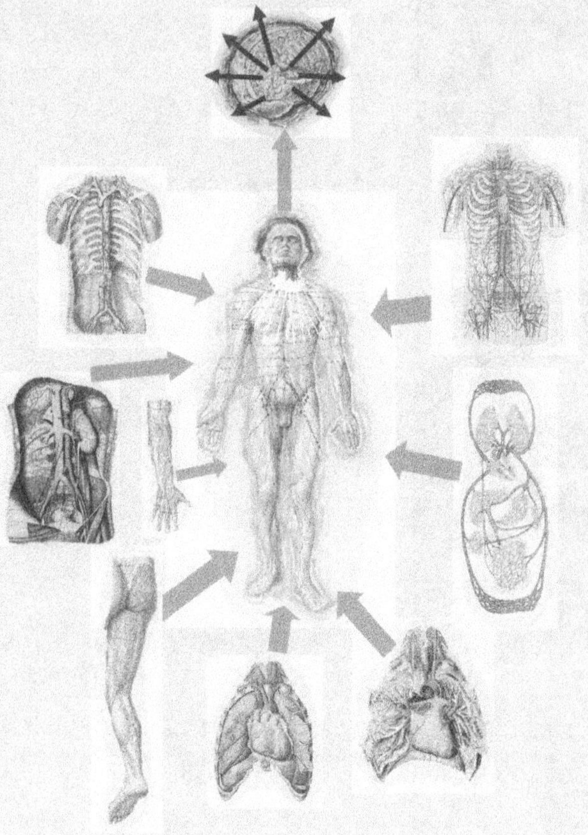

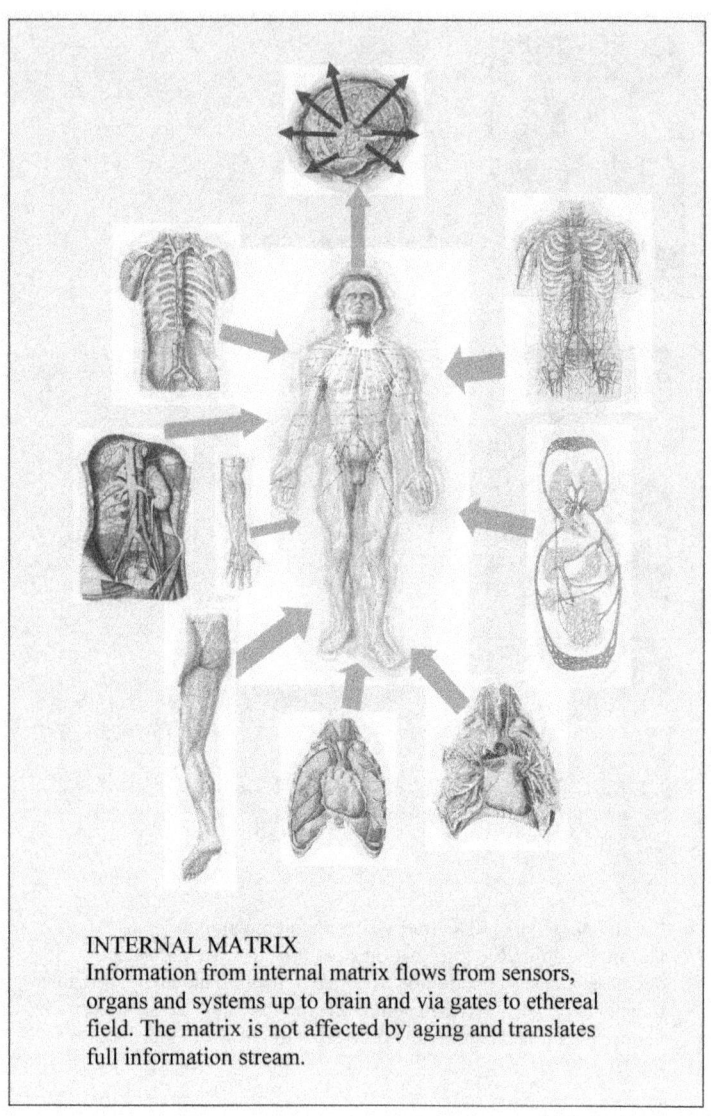

INTERNAL MATRIX
Information from internal matrix flows from sensors, organs and systems up to brain and via gates to ethereal field. The matrix is not affected by aging and translates full information stream.

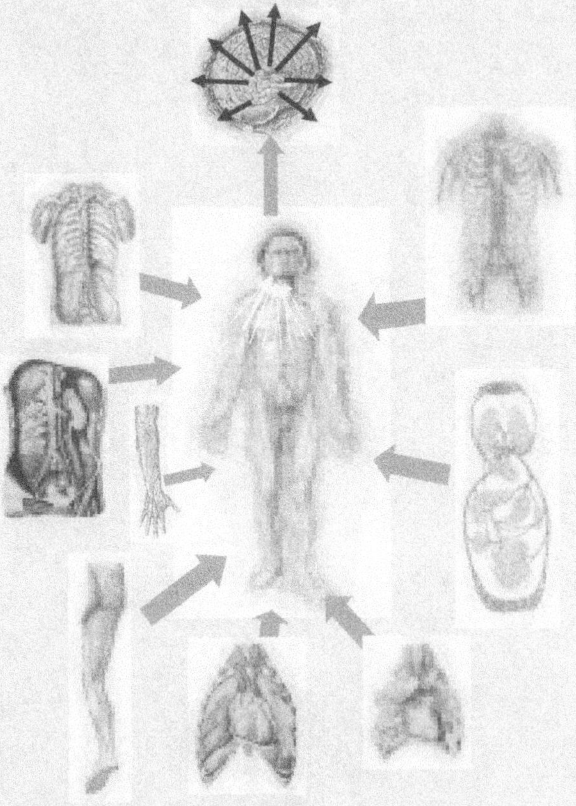

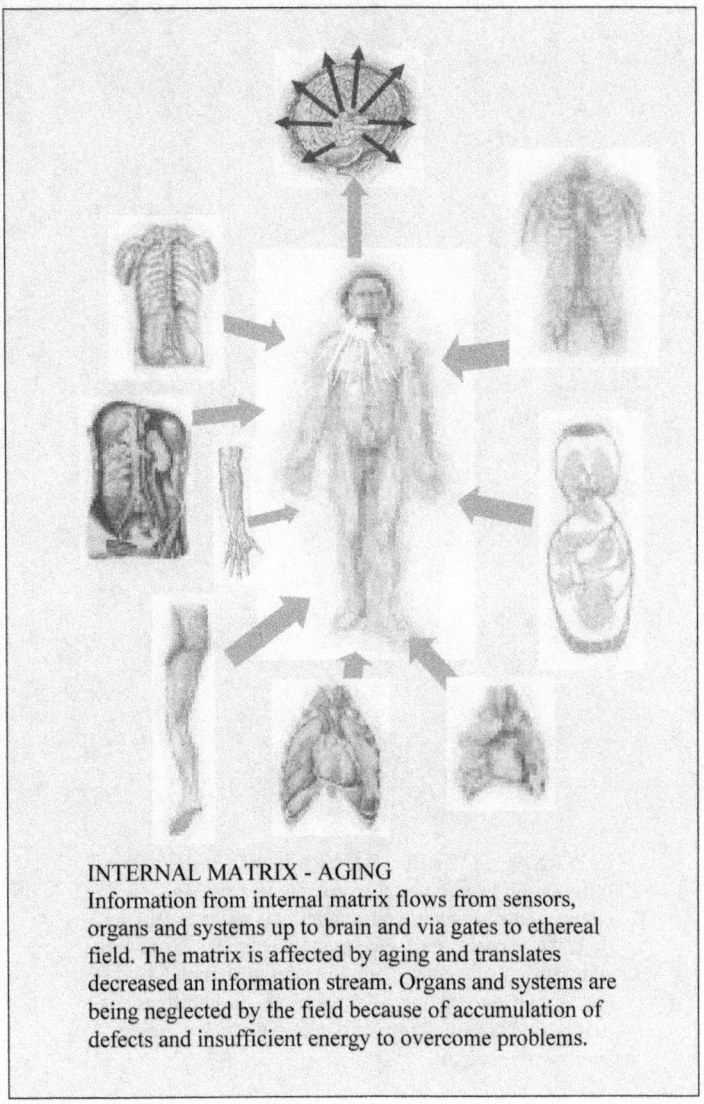

INTERNAL MATRIX - AGING
Information from internal matrix flows from sensors, organs and systems up to brain and via gates to ethereal field. The matrix is affected by aging and translates decreased an information stream. Organs and systems are being neglected by the field because of accumulation of defects and insufficient energy to overcome problems.

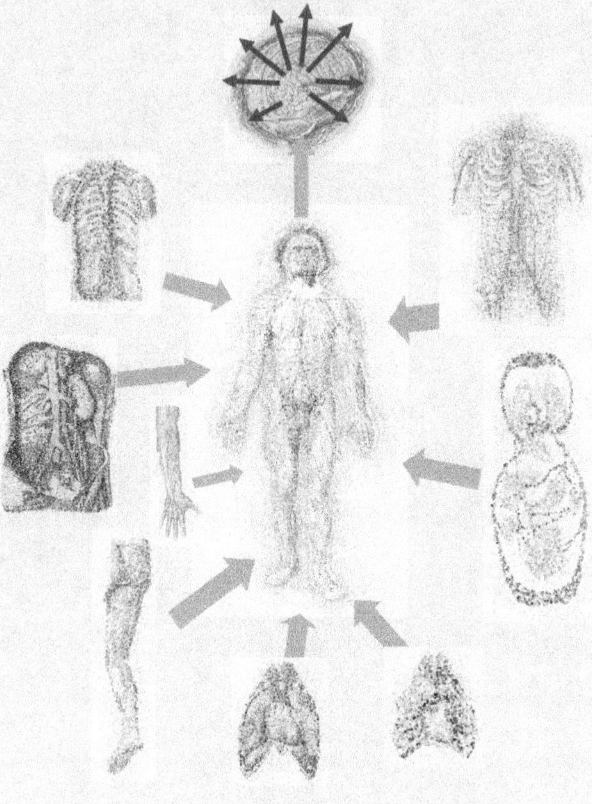

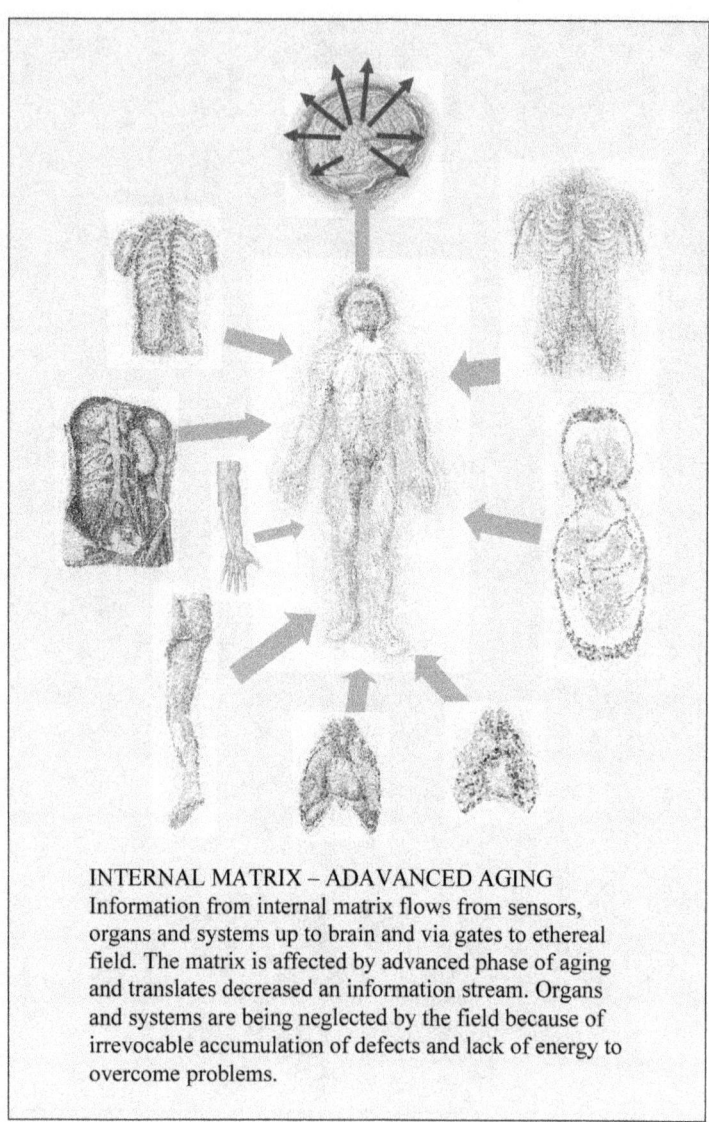

INTERNAL MATRIX – ADAVANCED AGING
Information from internal matrix flows from sensors, organs and systems up to brain and via gates to ethereal field. The matrix is affected by advanced phase of aging and translates decreased an information stream. Organs and systems are being neglected by the field because of irrevocable accumulation of defects and lack of energy to overcome problems.

THE EVOLUTION

The hypothesis reviews the evolution of the matter, besides the evolution of organisms is a part of all. As well, it reviews the process of evolution related to social, science and technology development. That classification is far from this to affect any contemporary theories and official concepts.

WHY EVOLUTION PROCESS

There is plenty of reasons evolution process to be required. The base of all is the origin conditions on the earth and the correlation between these conditions and the material world. From here follow the physical reasons that dictate development and impossibility to skip a stage. Other a reason is the necessity of precision transition from one phase to other, each stage to be accomplished before origin of the next. The process of evolution causes environmental change and is being influenced, and accelerated by it. Such change is heaping of biological mass as a soil layer.

Before to start talking about anything in that topic, let expound why the evolution is biological and whether there had been alternatives. That problem regards directly the physical bases of processes of development or eventually should be available an alternative?

As we have apprehended, there exists a media as the physicists from the nineteenth century have supposed calling it ether. That media has to be the conductor of electromagnetic energy. Later, the science, searching for real traces of that media and not finding denies its existence.

Our hypothesis categorically considers the opposite and retrieves the original supposing. The ethereal substance is basic from which expanded particles the condensed of the matter are produced. It is base of our matter and life, conducts information with a velocity much higher than the speed of light. The media allows hyper complex coded structures to be realized with distinguished functionalities. These elusive structures are independent or integrated to living organisms which operate by means of weak and fine electromagnetic impulses of hyper high tact frequency.

As we stated, ethereal media is featured as well by absence of motion of its particles, respectively, impulse of motion and energy. Being unable to provide, that media conducts energy for the processes which are controlled by the hyper fields. We called this system the Universal Energy and Information Conducting Media (UEICM). The energy is a basic parameter that determines the ability or impossibility of processes by given conditions. The lack or insufficiency of energy makes impossible creation and development of defined structures, but by the same conditions others should be available.

Electromagnetic fields, conducted by UEICM pierce the matter, including of the earth, but they do not possess the energy needed to origin large mechanical transformations of the matter. The only available transformation seems to be that of the particles, made through the whole earth's matter simultaneously, but slow, in to organic matter.

CREATION OF SPACE OBJECTS

First step to material world is producing of material particles which to build space objects. Everywhere in Universe are spread large space objects which, regardless of this, do rotate or not round their axis, have spherical or closed to spherical shape. Spherical shape is the one of the large celestial objects of plasma. Sake the hyper high pressure in their cores there are running permanent thermonuclear fusions and separating of massive quantities of energy. They are unable to cool and their matter is ever in a status of plasma. As the hypothesis asserts, the matter in the cores of the giant celestial objects is at the edge between primary substance and secondary one. They are energy sources emitting strong electromagnetic fields. Their gravity fields perturb the Universe far of their contours causing deformations in to the space.

Smallest sized objects, such as our planet do not support nuclear processes in their cores at that level and gradually are cooled. After cooling they could not change their shape any more, taking the shape acquired during the process. When up to their cooling have already obtained a spherical shape, such space objects save the shape, if no external influences, such as gravity or centrifugal fields did not made some diversion of it.

As it was expounded in the first part "Gravitation", the gravitation field round each free shaped space object aims to contract its longest measure, where largest gravity forces action. That process continues up to the moment of equalization of the measures of that body in the space (3D measures) directories. Geometrically that corresponds to obtaining of a spherical shape defined as geometrical area of points at equal distanced to one point, the center of sphere. When cooling is being accomplished before equaling of masses, mean, before obtaining of a spherical shape, the body takes that shape.

Why pay attention to the shape of the space objects? Because our life is on an almost spherical planet and large objects with other shapes are not being created in the Universe. That shape predetermines the conditions and features of our life. The uniqueness of our planet is its position towards the source of energy and physical parameters. As it was explained in the first part, a small part of the earth's semi condensed substance is being stretched outwards its geometrical measure by the concentration of its matter. That cover encircles the globe, forming a layer without a solid matter. That structure influences the geo conditions, because the gravity forces which action in are strong enough to keep the atmosphere closed to the earth's surface. In that way it makes possible and protects life.

As we have arranged our classification system, have extended the framework of evolution process out of its limits of life's timeline. We comprised and the physical processes in their development from the epoch of creating of our planet. In that way would consider the preliminary conditions and their influence in the context of the hypothesis.

The stages of evolution according the hypothesis do not correspond to these being officially apprehended. With this we do not engage anyone or anything from the official science statements, but only to make our narration clear. It concerns only the particular statements and conclusions of the hypothesis.

PHYSICAL PROCESSES

According the Hypothesis, the first stage of evolution on Earth is physical. In the framework of that process conditions life to be created are being originated. There are formed chemical elements and compounds. As consequence of the day and night cycle of rotating which corresponds to temperature differences begin processes of erosion of the cooled matter at the surface. As a result of the years cycle climatic zones are being distinguished, where the processes run in different ways. There is formed the spectrum of chemical elements and the preliminary synthesis of chemical compounds. All that is excellently known and we mention it here only in the context of our thesis. The first stage, the physical and chemical evolution does continue permanently. We should put a conditional limit up to the moment, when the physical and chemical processes approach to that stage, when life origin becomes possible. It is probable, and the Hypothesis does not consider this as its basic problem, at those stage simple organic compounds to be synthesized by one of the means supposed. Conditionally we would put an outline of that stage when the conditions in some point or area on earth become optimal the simplest organism to be created.

CREATION OF SIMPLE ORGANISMS

According the theories apprehended up to this time, life has been origin somehow, besides, the term "naturally" is being considered. But this does not explain how really it has occurred. Further, these theories claim, in Nature there is a tendency organisms to be created and as well, naturally, after billions of years of development that is also natural, to approach to the reason.

Unfortunately, there is no proof and rational explanation regarding the essence of that process that to make evident the ability of the matter and its particles being able to self-organize. The contemporary explorations of the micro world are impressive with their advance and show the presence of a large number of particles, charged or neutral, elementary or not, which behavior is obeyed only to the physical laws and regulations. From our aspect, behavior of some of them is strange, but there are no indications they to aim to self-organize in to an organic matter and at least, to origin some form of life. Further in the narration special attention would be paid to the introduced in the hypothesis concept for "Evolution chaos and order", mentioned and in the previous part, as being a logical argument.

At that location, being crucial for that material, the hypothesis claims, and life to be origin by the matter, formed upon the Earth's surface after the first conditional stage of development is unavailable without external influence.

According our concept, the Universe is being pierced by Space Information and Energy Field (SIEF), in to which are presented differentiated functional hyper fields. Physically their integration is based on hyper field structures. These elusive organisms could be highly intelligent creatures that do not have relation or integrity to material organisms and exist independently. By their peregrinations through Universe they explore each point and space. Being coded and provided with memorized information and algorithms regarding physical conditions by which origin of life is possible, they perform everywhere a comparative analyze of concrete conditions. There, where these conditions correspond to the required, origin of life is available.

If compare to our material world and our planet, where physical vacuum does not exist, in the same way does not exist in Universe a point or space, where life is available without life in to it. This is a result of the large number hyper fields that explore space and their hyper high velocity and penetrating ability.

There are theories, according which life on earth should be a result of conveyed by meteorites living organisms from other space sources. How these organisms should survive, knowing how bodies are being heated by the earth's atmosphere by their free falling seems to be a mystery. The explanation of the Hypothesis regards as well the opportunity similar organisms to be created and to exist at distanced points, worlds in Universe. Concrete conditions could content opportunity for origin of life, but not always for development and evolution process. That to be available extremely optimum conditions is being required. Otherwise, things finish there up to where should be able to develop and exist the living organisms. Hypothetically that means, somewhere by some conditions evolution process could be accomplished at too low level.

DESIGN OF LIVING ORGANISMS

The Hypothesis claims, so complex by its construction and behavior living organisms could not be created accidentally. They are being carefully and precisely planned and designed. Crucial condition to be realized is the opportunity their algorithmic structures – controlling fields to be synthesized and integrated to within the environment of ethereal media and the material world.

The second important condition is the possibility to be integrated to biological bodies that to live in material environment. As it is known, in the material media code bodies (processors and others) are synthesized on material, solid bases and create solid structures.

The choice of biological base is surely not accidental. Hardly somewhere should be some other form of life. From view point of our technology development a mechanical form is also available. Considering, whether origin an evolution of mechanical creatures is possible and relevant, or not, approach to conclusion that sounds not good. Biological structures possess advantages that exclude any of alternatives. They could be realized in an unlimited range of diversity, need little energy to exist, are flexible and compact. Biological structures are able to realize complex functional structures, such as memory, optical, acoustical and other kinds of sensors, conductors of fine electromagnetic fields and currencies with diverse frequencies, to copy algorithms of ethereal code structures and to self-reproduce.

Physically and even technologically, all moving parts by their movement are a subject of friction because ever one is supporting (bearing) the other. Friction leads to wearing out. To reduce friction, points of contact are being lubricated and the friction, respectively wearing essentially reduced. Biological structures in difference to mechanical content lubricant that precise comprise the parts and friction is being reduced up to slight. That reduces and the energy consumption by motion. Biological structures are able permanently to rehabilitate the worn out of tissue. Biologically designed functional organs allow realization of complex motion.

To follow that chain, would begin once again from the energy. Ethereal structures that should assist the origin of any processes possess low level of energy and hardly action on the matter that builds earth's mantle. They are able to penetrate through the matter without hindrance, but to action on the particles they need to concentrate energy in to a point or local space. One of the possibilities that seem mostly available is creating by the ethereal code fields of active fields using the energy of sources from material world, foremost, of sun.

The hypothesis should develop one of the possibilities the processes to origin. The probability to be exact and accurate is only hypothetical. After the conditional first stage of evolution the earth becomes a space object where the environment allows first biological structures to be created. This is ascertained by the pierced code fields explorers that provide comparative analyze. These bodies possess or additionally have added algorithms allowing them to create organic structures and first simple living organisms.

Probably, the first simple organisms have used an independent field as parent up to when they have been developed functionally to be able to self-reproduce. That corresponds to creation of controlling field that to operate the first biological organism. The possibility of controlling fields to penetrate the matter expounds availability of biological organisms everywhere, in all points and areas above earth's surface and immersed in the water basins.

The processes origin with explore of physical conditions and analyze of chemical composition. By presence of proper conditions, follows finding chemical elements and synthesis of simple organic compounds. The object is to prepare the building components for the biological mass of first simple organisms. That is achieved by further complexion of synthesized organic compounds. A second object is, to make ready organic or not organic mass to support life of the said organisms, their metabolism. During the process, the first organic compounds are being disintegrated, but create preconditions more complex synthesis and future more complex biological structures. Insufficient energy of controlling fields that provide these processes does not allow to make activities at once, but spread in the time.

Basic regulation of the evolution process is careful and precise planning of each stage to support and provide creation and development of the next. Creation of simplest organisms is the only available by the initial stages of evolution, when conditions for life of more complex ones do not exist. Creation of more complex species is blocked as well by the lack of building material. The lack of energy necessitates smooth transition from a stage to a stage. Problems from the side of controlling fields and integrated codes do not exist. Problems are from the side of material world, where processes need energy, are inert and it takes too long time to achieve a slight progress and to accumulate biological mass. The new created biological organisms need to support their metabolic processes, for that purpose they utilize and the biological mass of these, being created beforehand. Utilize of readymade biological mass of other organisms is one of basic principles in material world. Without that, complexion of organisms is unavailable. In that way, the organisms do not use chemical elements or simple compounds, but complex organic ones, from where extract what they need and save time and energy. Other ways, metabolic process should last long time and costs a lot of energy. Gradually, organic mass is being heaped in environment, basically soil, to be building material and to support life.

Each stage of evolution is being achieved by proper intervention in the code system of controlling field of the preceding. The codes of new species are based on existed by modification of these that would correspond to the new species. Creation of animal and plant living organisms is being complicated and their diversity increased. From one side this is an objective necessity, because ascending in hierarchy of species requires more complex organic mass to support metabolism. These requirements are related to more complex chemical compounds to ingredient the food of organisms.

To that stage we would look at a view point that consider as being important. At that stage is being formed the intermediate link that could be signified as transition to implantation of reason on earth. What is that intermediate link and how we should be sure, it is really? Logically, implantation of reason on earth seems not to be by chance, so as nothing is occasional in the endless rows of species living on earth's surface or in submarine world, flying in skies. Implantation of reason is logical in one case only – when mind is in synchrony with perfect universal abilities of motion and access to everywhere on earth. Creature that would carry that reason must be able to utilize it. We do not imagine a fish with mind, even only in water; it could not utilize it rationally.

In the great chain of evolution are being outlined monkeys and apes. They are not by chance falling in the field of vision of Darwin. Monkey has a basic supremacy over all others. It has excellent developed functions of motion, but the very important is its function "catch". Supposing, it is derived and developed from lemurs million years before. Apes have already formed hands and could make use of them. Function "catch" have and others, but they do not have the properties of the apes to feel good everywhere, to be so moveable, and finally, to be able to content and utilize reason.

We could say, during evolution there has been a purpose – organism to be created, creature that to correspond to defined conditions and to be higher in the hierarchy of species for being able to acquire reason and to be useful with its possession.

Each stage of evolution consists of numerous sub stages which correspond to creation and development of each new species. According the hypothesis the entire process was being planned carefully and precisely, in to it is implemented hyper intellect, unlimited patience and exclusive sense for harmony. Precisely are accounted all initial conditions of environment and their further change. These conditions include geometrical measures, rotation round own axis and round sun, tilt of the axis, chemical and physical conditions, climatic changes, geographical position, cyclic condition changes. Each change of species, and arrange to given conditions, creation of new species is in a relation with controlling fields and implemented in to it common features of that species. By all cases there is a standard formula that defines specie's general features. In that standard formula are previously foreseen diversions, related to changes of conditions of life.

Each controlling field operates the integrated to it biological organism besides it sets in order its micro and macro structures, organs and systems in to space and according their functionalities. At the beginning of some diversion, for instance in living conditions, the algorithms start to arrange biological structures by manner corresponding to that deviation. In some cases these diversions had been foreseen, but in some – not.

Then, a species obtains change besides it adapts or disappears, but in some cases a new species is being created based on the given. The standard code of the new species has been changed, compared to the previous and it initiates to develop so as the new or changed algorithms would require. By the inertness of processes in the material world, particularly biological, for the new biological organism which corresponds to new species to approach to the requirements of the new standard is a long lasting process. It is impossible to be achieved in the term of one only generation. In the framework of the process each next generation adds new features to its and in that way gets closed to the new standard. Finally, species approach in their biological essence the standard requirements and stop development, if new or changed algorithms do not require it.

The ideas and meaning of origin and development of so much species is a mystery at all. But, it could be expounded with harmony in Nature and necessity of various biological mass for the organisms to support metabolism. Their dependence on other species leads to creation of a large ecology system.

HARMONY IN LIVING AND NON LIVING NATURE

The precision of the evolution process is being performed stupendous. By planning of that process physical and geometrical factors specific for earth are being a subject of special attention. These parameters define the features of the habitants, including their size and appearance.

First of these factors is the size of the earth, defined by planet's diameter. The aim by planning organisms is they to correspond in adequate harmony to that measure. The size of each plant, animal and foremost, of humans are being considered with the planet's size, besides, the habitants must feel in comfort. In that way all static and dynamic strength problems that should occur are being reduced to minimum. Respectively, is being reduced energy consumption and biological mass, needed to support metabolism.

The living organisms are designed in a harmony between themselves as a consequence of their harmony towards the earth. The size and features of the living creatures are optimal and toward the physical parameters of the environment. Such are the gravitation field and forces that load the construction of each organism, the atmospheric pressure, consequence of gravity field, climate, presence of energy and more. An optimally designed organism should experience a minimum of these influences. Bones and organs of a tall and heavy creature should be extremely loaded, as a result, deformed, bended or break. To make them strength must design to be thick, what means, more heavy. More heavy brings more problems, including need of more food, living space and energy.

In that chain of development, adapting and forming some diversions are being presented. The most distinctive is the epoch of dinosaurs. They have sizes quite distinguished and unusual towards these of our present times. Has been that stage of evolution extremely necessitate or simply an experiment to approach the optimal measures? We should guess, considering, them as being a large biological mass, needed to heap organic matter on earth for further evolution processes. The topic regarding the dinosaurs is very popular; this is a mysterious period of the earth's history and evolution that has been irrevocably gone away. The interest is being fomented by their unusual measures, appearance and features that we could know only from fossils. The theories related to their death sound like that – giant creatures should be destroyed only by a giant strike on them. Such should be foremost a falling meteorite that sweeps away and destroys all on earth, including dinosaurs.

In the context of our theory other a probability appears. Let to the meteorite add one more, biological reason, namely, total and fatal disease that to destroy them. Monitor lizards that are survived have in possession a reach collection of viruses to which they have been accommodated and live without problems. Regarding meteorite would say, if such catastrophe destroyed the dinosaurs, it must destroy and others. The consequences of such a catastrophe evidently should not be selective – strike only on dinosaur's heads sounds not realistic. If other has survived dinosaurs as well, a number of them should be living in our epoch.

Dinosaurs appear in an earliest stage of evolution and we would not consider whether their appearance should be extremely necessary. For more than 140 million years, dinosaurs reigned as the dominant animals on land. What is evidently, their giant size does not harmonize with planet's size and environment. When we consider evolution process as being carefully planned, sometimes occurred these animals have been bigger than necessary, do not harmonize to environment, need a lot of food, are predators. They lived too long, but seems, at given moment become to aggravate and perplex the further evolution process. As tough, 140 million years the process has been neglected, leaving these giant predators to predominate upon earth. In an instant they become incompatible with future evolution plans, they are no more beneficially link if have been before. Future evolution plans require origin and development of creatures that to harmonize with environment which to produce that one, able to carry a reason. Dinosaurs could only harm such a development, being too far from the trend. We know well the problems caused by crocodiles, but these problems are local. Let imagine the picture to live with dinosaurs in our epoch what a global problem should cause. If they should be preserved, today's civilization could not exist. We are in the way to suppose one more reason dinosaurs to be destroyed – they become noxious to evolution and must be eliminated from the scene. Destroying common codes, similar algorithmic systems, specific only for the group of dinosaurs, could be a probable way to discontinue their existence. That is being achieved selectively, without influence on other organisms. The Hypothesis researches dinosaur's destruction in the context of a serious reason. In a stage of evolution it occurred they restrict further development and must be removed.

As other distinctiveness would mention presence of organisms, origin and existed by unusual conditions. They are foremost these, being able to live by extreme temperatures, closed to polar or equatorial zones, by extreme pressure deep submarine, subterranean inhabitants. Their creation is a result of the ability functional hyper fields to penetrate everywhere through the matter and to create biological organisms everywhere it should be least physical possibility. They are provided by algorithms to be able to design, origin and develop such organisms that to live and support metabolism by extreme high pressure and lack of light and oxygen, by high or low temperatures and lack of water or food.

EVOLUTION CHAOS AND EVOLUTION ORDER

We include that chapter before to approach the human as being the most important link of the evolution. The concept "Evolution order and chaos" was introduced in second part, "Life" with a stipulation to be detailed expounded in current part. That is a method of abstract and deductive considering by means of which we make attempt to support the claims of the hypothesis. We hope it to be a strong argument in our train of reasoning. The method is based on the probability something or some event to be possible. On contrary, thing, fact or event that really exists to exclude the first claim as being impossible.

The official concepts regarding origin and development of life are pure materialistic and do not permit any external influence. On the other hand the complexity of biological codes of DNA requires some complex of logic and design to be created. Contemporary theories claim, all that is a product of material, natural evolution. During billions of years, step by step such complex structures are being possible to be realized.

For a while we should accept such a claim as being correct. That means, sometime the particles of material world somehow have been organized into organic compounds and in that way created building material for organisms. Creation and heaping of organic matter is probable, but it is non-stabile and complex compounds trend to decompose to simple ones. But, let accept, organic compounds are being created and grouped in to a simple organism.

By condition that process being random and not controlled by external force, it should be as well not planned. Being not planned, it should continue in each moment permanently. That is to say, in every moment organic matter and organisms to be created and heaped. In that case we should find between the species intermediate stages. But, we do not perceive anything like this. All attempts to find Loch Ness monster or snow men did not lead to success. Even to find snow men, it should be a single exception and not the core for our concept. Our concept is based on this to find out some reason logic by these complex processes having in mind the beginning is billions of years before. And what the result might be if there is no logic, no intelligent design as the official concepts proclaim. Might be the result a total chaos?

What we could mention in that short while of our presence on earth are clear, entirely developed species and no intermediates. According material development and absence of plan and external influence, round us must be primitives, all stages from ape to human. What we identify is but only monkeys and humans, all others we find accidentally as fossils. As well, we do not mention any development of species – lemurs and apes and all others are the same as have been million years before.

That considering is applied at the timeline. All here is conditionally, does not affect exact time and epochs being only a principle illustration.

At the timeline proposed two open lines are shown – the path of real evolution between ape and human and the line of apes that never would be more than apes. If evolution process is material, not planned and not influenced externally, there is a chance at each moment an ape to take its way of development. The closed lines represent what should be if that was real. Currently we should be living among and contemporary of all kinds of semi humans – what we called “evolution chaos”. Do we?

The Hypothesis is specific – each species in chronology and hierarchy is being developed on base of the previous by means of some intervention and change of its algorithms. That has been made not only systematic, but at once, accomplished, once forever.

Within the code systems of the species is being loaded development, but in the context of expected changes of conditions of life. When these changes of conditions exceed expected and previously coded or become other influences, simply, a new species is being created, instead infinitely to change the previous one.

MAN – THE CROWN OF THE EVOLUTION

According the hypothesis the so called evolution is not a spontaneous process but product of intelligent design. The evolution is nothing more but development of the species following the development of the earth's conditions and the inertness of the matter.

The so called evolution consists of different stages or epochs when new species are created. The direction is apparently from simple to complex and the object of this process creation of species possessing wit.

It is evidently, that stage is not dinosaur's epoch, because they should break the entire development. To approach that stage, a creature must exist that to become grandparent of the reason bearer. That creature must response to defined conditions that already were described. By all of cases the reason bearer should not be a reasonable, but helpless in its actions and motion, or to stay predator, and so on. In the theory of Darwin that creature is monkey, namely, ape. Really, by its appearance and features ape could be considered as being the grandparent of human.

There appear two basic concepts for human creation and development up to day. According material theories, the monkeys or apes in a stage of their life are leaved the jungle, respectively, descended of trees, and become living by diverse conditions – in savannah. Changing their manner of live, they are being compelled to adapt. During a long lasting process of adaptation apes begin to work and become a reason. Their development is going slowly, but some sixty thousand years before suddenly something happened and they accelerated development. The theory that expounds how and why that occurred is based on a giant volcano in Indonesia that destroyed most of the humanlike creatures and a few of them survived. The consequence is a quick development.

The questions that we have in that context are similar as by dinosaurs. Is volcano acting selectively? Why there are not undeveloped humanlike creatures, but all have been influenced equally? There are plenty of questions; instead to ask we proceed with our claims.

The hypothesis is being abstracted from all details of elapsed millions years of evolution to approach today's human reason and appearance. The line that we follow starts from revealed above in current material and passé through the questions why and how. We are talking about precision and planned evolution. What is puzzling is the time elapsed from the first signs of transition from ape to human of some millions of years. As tough, this seems to support the theories of purely material evolution by chance. If evolution is pure material, it should pass exactly so, during millions of years, heaping slight changes. Seems fully realistic the claim, conditions are being contributed to achieve that change. A planned process should be accelerated, besides changes to occur more rapidly.

On purpose to initiate explanation of its concept, the hypothesis sets off directly towards eventual object of a planned evolution. We stated no one makes something accidentally, still more, the Supreme Reason. The object of evolution process is a creature that to possess all features and qualities to bear a reason. As well that creature should not be any crawling and helpless species or that being unavailable to leave the water environment. The bearer of reason is the world's master. To his features must add exteriors like aesthetical appearance and harmony of body's components. As well to add internal qualities for which exterior is interpreter. These new qualities correspond to the new status of reasonable and creative creature. Gradually, these qualities are being implemented and during development the human approaches to that stage when is ready to solve the problems for which he has been assigned, being as well and final objects of evolution. In all that we perceive the harmony that already described.

The precision by human creation is varied and does not accomplish with his appearance. Physically, he must be the most perfect creature in the world. His measures have to correspond to all that exists and to the earth's size. From the view point of contemporary technology development, these qualities are crucial, should we imagine a modern jet and creatures appeared like gorillas on it

There is an important factor that we do not have to omit, when talk about physical principles and reasons.

According the General Theory of Relativity all is relatively, including the time that elapses differently by distinguished conditions, by low, conventional velocity of motion and by velocity of light. Upon planet earth we are used to our concepts for time, being relatively to continuance of our life and earth's rotations.

Comparing to that we account a given period as being long or short. But, when we have in mind an in principle distinguished Media in to which we, our planet, the material part of the Universe are being immersed and which media has originated the material worlds. That media is also material, but of quite different physical parameters. The hypothesis claims, the velocity of conducting in that media could be hyper high, much more that velocity of light in material world, the same, to compare the snail and electronic mails. But, let emphasize again, the speed of light is restricted by its emitting by the inert particles of our matter.

The exclusive physical parameters of ethereal media are related and to conception for time elapsing. As it was revealed, the particles of ethereal Media are of expanded matter, have no motion, impulse of motion and energy. In that way the concept time as far as it is related to motion does not exist. Of course that is not an obstacle the fields integrated to biological organisms to elapse time and cycles of our planet as precious as we do. In that way, these controlling fields regulate growing, development, daily existence and aging of the integrated to them biological organisms according earth's cycles.

In that context we elapse the time running for the distinguished periods of evolution through our criteria.

From aspect of the continuance of our life, from the time when first organisms are being originated up today a long period has been elapsed. From aspect of infinity our life is infinitely small quantity, reciprocal to the infinite. From a view point of ethereal media that term is so long as it is necessity, elapsed by earth's criterion and does not have a meaning if elapsed by measures of the ethereal media.

Why there has not been performed acceleration by transition to human? Difficult question, but to try to approach to the true seems not entirely impossible. Surely, all theories of adaptation to environment are absolutely true. Each modified genetically species is being leaved to be arranged to environment and environmental conditions changes for long years. During that period runs as well and peripheral evolution process, climatic changes occur. In principle, each adoption is possible only when in the algorithmic system of controlling field is preliminary coded such opportunity. On contrary, the species sticks rigidly to its standard being unable to be modified. By a lack of codes that to correspond to tolerances of adaptation, the species should either survive or disappear. But, this is not the essence of the current considering. The essence is how a reason is being acquired by a struggle to survive; besides not only the species is being changed, but its mental abilities radically.

The base is the basic species to approach that level of development to become basic to acquire a reason. It is not occasional belonging of all chain up to human to one ancient species – lemur. The common is function "catch". Function "catch" is further being developed by gorillas and especially by human with weakened clutch, but with increased sensitiveness and skillfulness. Let keep in mind, to human in its long lasting development is forthcoming to assimilate the experience of processing the material environment, besides the period of manual work is incomparably longer that the one of machine processing.

MISSED LINKS - IS THIS A PROBLEM

Searching the link between apes and human that to be closed mostly to apes should bring us a proof, we are origin from monkeys. But, to find that missed link is a problem, not because it does not exists at all, but because the fossils of that mysterious semi human are somewhere deeply in soil, or decayed already. Search of missed link is a material proof necessity for the material theories. Our hypothesis is not interesting at all what is the skull that has been found, is it true or an ape which teeth are being rasped to the size of supposed exemplar. We simply do not interest about that missed link at all, because our method is logical and we make easily an interpolation between the stages in the same way.

TIMELINE

The only line of evolution is between ape and Homo sapiens.
It has been originated once during the time and never more.

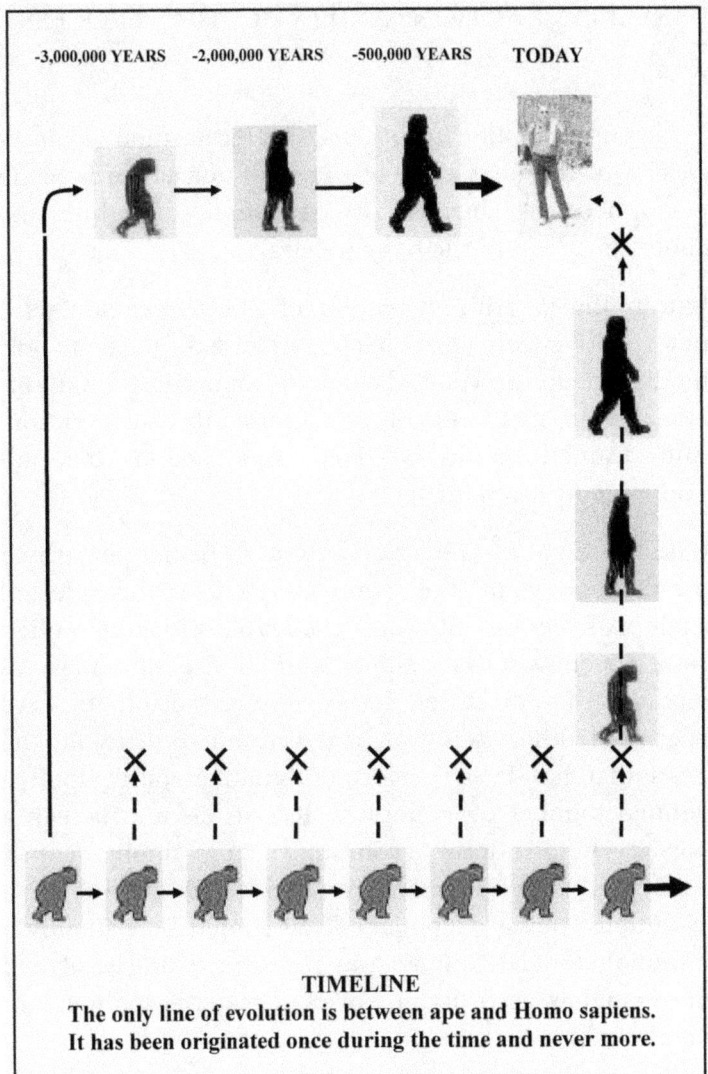

INVARIANT ACTIONS AND EVOLUTION PROCESS

According the hypothesis, the transition from a species to the next stage of the evolution chain is being performed by means of intervention within the controlling field of the basic species.

Within the algorithmic system of selected exemplars a new code system had been implanted, at first only modifying the previous. There are numerous of variants available of methodic of performing the intervention. But, distinctly could be fixed these actions that are undoubtedly invariant.

First of them is, the intervention is being performed simultaneously in the controlling fields of selected exemplars, besides all others are leaved with unmodified code systems. It is possible later, if the first wave of mutation has produced waste or unsuccessful, to have been and others waves of intervention. But, by all the cases that has been performed simultaneously upon a defined number of exemplars. If that was not in such a way it should occur evolution chaos, simultaneously existence of developed at various stages of the same evolution process exemplars. That should be evidently if evolution towards human was performed not as planed but somehow, according theories of material spontaneous development.

A direct consequence of that claim is whatever impossibility from the rest of the apes, these, that are not being object of influence, by whatever conditions to become humans. More, all desperate experiments to teach them to speak or to obtain some level of reason give no results and are doomed to failure. There is a very simple reason – in the fields integrated to these exemplars never has been modified codes that to correspond to some transition. Including, their speaking formula has never been modified, simply, does not exist the corresponded modification performed by humans during long lasting period, etc. The top of opportunities of all existed monkeys and apes is their increased communicativeness and ability to copy and reproduce human's actions. That is due their long evolution and preparation at the level to become able to generate the humans. In that chain they have obtained some intellect at some level that is but still animal's one.

The second invariant action is related to the original algorithm which we called "standard". "Standard" is a system of codes related to given organism being implanted in its controlling field. Standard defines initial conditions of origin features of given species.

To standard codes ever a range of additional algorithms is being added to meet future environmental and other changes. That range contains preliminary defined and foreseen tolerance of presumed future modification. These modifications are diversions from standard allowing to the species to adapt to environmental and other changes.

By vice versa change the species returns back to standard, or modifies in other diapason defined and previously coded. The essence of the second invariant action is the standard algorithm to be absolutely equal for all selected exemplars in which is being implanted. That means, all exemplars of apes, selected and objects of intervention acquire simultaneously similar new standard. In that way during the next periods of their development they have equal initial opportunities and equal aspiration to approach to that standard.

From aspect of origin and initiation of humans, in context of all race theories, race differences absolutely do not exist, all have implemented equal standard features. Differences between humans such as skin color, appearance etc. are a result of diverse conditions like geographical position and environment, climate, work and opportunities to find means of life and food.

The hypothesis states clearly and unequivocally - any differences between humans and their mental opportunities are excluded namely sake of that second invariant action, by which all have had obtained similar algorithmic modification. During the process of infinite migrations people have changed their conditions of life. By these migrations they have searched the best conditions available – climate, water, soil, food.

HARMONY AND PROBLEMS OF THE EVOLUTION

It is indisputable the will of our Creator, the Supreme Reason to arrange the earth's life according the principles of harmony, good relations between the habitants. Of course a lot of people do not believe in this, especially the theorists of the contemporary materialistic concepts.

For the people the question is about ethics, moral, and absence of hatred, envy and malice. Why these good wishes are impossible and ever there is something negative that to compromise their implementation? Foremost, the opportunity of intervention from media where there is no energy requires transform the energy of the material world. Namely, the first problem is energy. It leads behind a number of additional concrete sub problems. The energy for biological processes must be economized.

Second problem comes from the side of material world. This is the inertia of the matter. Slow processes in material environment require the organisms to utilize readymade organic compounds as food for their surviving. Synthesis of simple compounds should be long lasting and the organisms spend their time only swallowing. That problem is enhanced by supreme organisms that need the biological mass of other living organisms. It is not necessary to comment what that means.

Third problem is surviving of species and reproducing of health and viable generation. The principle is well known – survives the strong and health. To that entire let add the social life by species, foremost, by humans.

In these three problems are loaded the cruelness, the predestination of some species to be food for others, rivalry and many more. In modern life to that let add the aim to live in luxury and the wish to benefit all advantages of civilization.

All that shows – life on earth to be organized is an act extremely complex. Despite the Great Reason who has created it, precision and aspiration to harmony, there exist problems that surely have been foreseen and such that have not been possible to foreseen. The same is when create something, but do not know its behavior after a time.

In that context should take a look at the science that tries to take part in the evolution process. The hypothesis does not refuse that, supposing it should be its acceleration and probably is a normal to be considered as a stage of it. The earth's reason being closed to earth's problems as normal is trying to solve them. But, on the other side of balance is billion years evolution some of which results we try to modify in a few years. To do that, surely there are and risks because no one could predict what should be the further behavior of the organisms mutants created by us. No one could say what should be the influence of the genetically modified organisms on people and nature not during our life, but after generations. Should we pay for the violence upon the nature? Is this similar, to have a new car and to replace what just as the fancy takes until it acquire the view of functionality that we would like?

The primitives from which evidently origin humans are being saved in the same view as they have been in the dawn of human's development, millions of years before. Of course these, who have been survived, namely, to which the described already tolerances of code systems have succeeded to preserve them by the various diverse of life's conditions during so long period. Lemurs and today are lemurs, monkeys and apes are the same monkeys and apes that have been at that epoch. It is out of question about any "natural" development. If such was in fact, then lemurs, monkeys and apes should move somewhere further in their development. No, they did not.

How the hypothesis evaluates appearance and foremost the contemporary human's development. That material is based on physical principles and by that analyze us would start from there. What has been achieved with human's appearance? We would return to the energy problem. What human is able to do no action from ethereal media should be able to perform in the same way sake of insufficient energy.

All organisms from evolution hierarchy up to human could only to reproduce themselves to infinity, or till conditions allow, or to perform the same actions or behavior. In difference to all, human is actively acting in the material media. He is able to activate the energy of material media to create what for controlling fields is a problem. Regardless, the Space Information Field contents information compared to which ours in still too insignificant, it needs human as a mediator in the material world.

The function of humans is to utilize the energy of the material world and to realize technology advantage as being a part of the process of development. In its development, the Project has created the human as being irreproachably harmonized to the size of planet, habitants and other environment. That creature possesses sharply distinguished appearance compared to all others, skills, knowledge and talents. Human is able to get in any material media, possesses perfect mobility, abilities to explore and acquire knowledge.

If evolution is being planned, there are evidently defined purposes that should not be performed directly from the side of ethereal media and the means to be done is human.

ACCELLERATED DEVELOPMENT

We have the chance to live in a period of accelerated evolution. Today changes come faster than yesterday, tomorrow would be faster than today. Official science is being in defined situation of confrontation to religion, considering, nothing should exist outside the material world, doing explorations and experiments only in that media. The confrontation comes, as far as the religion talks about something spiritual or out of our material world and our hard imaginations. Once again would like to repeat, surely at least one scientist skeptic in the world on Sunday visits the church, but on Monday returns to his material concepts that evidently are not in unison with the Sunday's sermon and prays.

Before to take up with that part of analyze, the hypothesis sticks up to declare its honor to all religions that respect the living Got. That is not a gesture, but a consequence of all said up to that moment – we have One Creator, regardless, how we accept or call and with which rituals honor. The hypothesis is being created after careful read the Bible on and between lines. Claims and conclusions are not accidentally and should not be thrown away and neglected quite easily.

We return to the period of accelerated development, contemporaries of which we are. From all conditions on the earth, from all variants and opportunities there is one geographic position with best conditions. No extremely temperatures, soft climate, friendly environment, all that needed to give to humans the opportunity to accelerate their development. In to that geographical position is concentrated mass of humans. But, there is one more condition that is required, the consciousness of these people. The question is about, which society and civilization are being survived during the centuries and why. Evidently, there are not civilizations that are being heathens and that make a bow to their mythical gods. To be converted to the living God is the most important condition that new age of evolution to be started. Then appears The Savior of that society and with His birth namely that new stage is originated. The Savior is the link between creators of Renaissance, technical and technology revolution and God.

Here the Hypothesis would restrict not to explain what is round of us and what all we know well. But, would like to return to the origin point of the biological development from which millions of years are passed. Has it been impossible to start directly from here? What should be the meaning of a silicone world without our harmonic natural environment? On the other hand, is a planet with developed pure technology civilizations, without living organisms probable? There was expounded why not– the fields from ethereal media does not have in possession energy and tools to do that. The living organisms are mediators the energy of material worlds to be mastered and reasonable a tool to transform the material environment.

The contemporary stage of development we should call "Silicone evolution". Entering deeper in the micro and Nano world brings new discoveries and new opportunities for science and technology. Are there restrictions and which they should be? First, we should analyze some hitherto restrictions. Such is the internal combustion engine. All of them are obeyed to the thermodynamically cycles of Otto and Diesel. They are physical laws and nothing could be made against. Already 120 years these motors are based on the same physical principles, regardless the exclusive technology advantage which could only improve their durability, mechanical balance, noise and vibrations, to decrease fuel consumption. The cycles limit a principle development of internal combustion engines.

Other an example is in location to build a supersonic passenger aircraft. There are plenty of aerodynamic restrictions that have done "Concord" to be less economical related to conventional aircrafts. Attempts to build bigger one meet strong restrictions of the same character that make it unavailable by contemporary level of technology.

It is probable in the process of exploring of micro world to approach a stage of saturation, what means, to make the next step to require building of more and more expensive and inaccessible devices. It is probable idle in each location of exploration and technologies, besides one of them to disadvantage from the others. It is probable standstill by increasing the tact frequency of processors and memories from all kinds. We should imagine the development a winding line that approaches in a moment its range of standstill. That means, need of more energy, resources, time to achieve what has had been achieved by less efforts before. It is probable, the physical laws and regulations to lead to limits.

The hypothesis is far from will to be a bad prophet, regardless, there exist some physical restrictions. They are well known, but now are being included in the context of a quite distinguished analyze and basis to compare.

First is inertia of the material particles, respectively, of all the built technology works. From here follows the need of more energy. The second restriction is the speed of light. It is direct consequence from the first. That limit is a result of inertia of the particle that emits electromagnetic impulses, respectively, light. Electron that by its oscillation emanates these impulses is a material particle from condensed matter with inertia that limits its tact frequency. The speed of light is limit for the space flights that should take long years to approach the nearest neighbor systems. The speed of light even to be approached makes us mostly attached to our world and sun system.

The third restriction is the necessity to implement integrated code systems on hard bearer, on stuff. At that stage, when no one is aware in the existence of the other media, or only foggy explanations and guesswork are trying to pick up the end of the curtain, to integrate in to it is inconceivability. If someday the science commences to ponder where the annihilated particles are going, and even accept the existence to perform integration in that media, hardly humans should be able to create integrated code bodies in to it.

Hypothesis supposes a strong restriction at the boundary between both media should occur. From our time could say ethereal Media is not yield to be explored from our material world. The impossibility to integrate in that media makes the aspirations a living organism to be synthesized fully not probable. The alternative to combine integrated on hard material code system to synthesize by our conditions living organism sounds grotesque, if possible at all. Let imagine a virus created in laboratory and attached to it processor, regardless, how miniature and perfect to be.

Regardless of these restrictions expected, the science has unlimited compared to our considering opportunities in an exclusively wide diapason of actions. The science will go ahead and this is its purposes and these of the accelerated development.

HUMAN'S CONTROLLING FIELD

We said, within the algorithms of a long consecution of species systematic discreet intervention had been performed their features to be modified. The end goal is to approach to creature enough perfect that to be able to keep a reason and to utilize it effectively. All other species are being stagnated at their initial stage of development, because no more interventions are made to them. The idea is: no evolution and no development is possible without external intervention and self-development is unavailable.

The code systems of human contents information from the simplest organism, from the beginning of the evolution, of all stages through which have been passed and modified information related to the diverse from the animal's species. Finally, the code system of human contents the new information that is quite distinguished in its principles compared to all that had been before. It is evidently, to skip a stage is impossible, even the one that brings the slightest modification and to go directly to highest. Therefore, the independently human development is being lasted as well millions of years. It is acceptable, by these modifications waves of modified codes to be implemented and further - corrected. But, there is implemented and a wide diapason of probable and admissible diversions by development to make the new species to survive first and to develop, second.

We would accept, there is implemented a standard and after the intervention the exemplars selected and object of it are being leaved to save themselves somehow for too long period. Wee said as well, the algorithmic fields of third type are able to communicate to SIF (The Space Information Field) and to acquire from there algorithms that to provide them with new abilities, such as talents, skills, to acquire and analyze information. That was described in part two. Here we should remind the Spirits Santos, Solomon, David, Alexander the Great, great Leonardo, Newton and many more. We have said the code systems of humans are equal in rights towards the standard implemented. But, people are not equal regarding skills, talents, memory, activity etc. and we consider that as being normal. Only distinguished individuals acquire an additional algorithmic supplement and information and that is not accidentally. Besides, that intervention is in most cases in harm of other codes and some of these individuals have problems with health status or behavior. Let remind blind and deaf genial musicians. Other a fact is making conspicuous and probable, these interventions are not inherited, surely, not in that way.

In difference to the second part, where controlling field of human was described here we would make that in context of the development. During the process of development there is accumulation of information and corresponding forming of biological organisms. There is not a stage that to be skipped and surely that is unavailable. In that way we could give an answer to the question how human has been created. According the hypothesis, he is created, but not at once, besides, his code system includes information that is compilation of all previous. No one has made these codes no more again because they already exist and that is unduly. They build corresponding biological structures in the organism being followed and explored by the science. These building particles and their codes are remains from passed epochs.

The object of the current explanation is to emphasize that nothing has been built on nothing, but on carefully and precision create and developed basis. Therefore surely are particular physical reasons. Foremost that is again energy, second – the necessity experience to be gained in earth's environment and in conditions of biological world, training of biological organisms to survive and adapt. During the millions of years the organisms have gained these features and built habits, skills, motion and other functions, ability to observe and analyze. Could be human created from nothing? Surely that has not been achieved and is a heavy problem. To do that, a large code structure charged with giant concentration of energy should be needed. There is needed building material that to be synthesized at once and from it to create and assemble a human.

The evolution reveals two basic methods of synthesis of living organism. The first is by external integrated code fields, especially designed to do that. Supposing in that way quick synthesis of simplest organisms is available and we could be aware in a relatively high level of certainty that it has been namely so. After their creation by external intervene these simple organisms already have own reproductive codes and corresponded biological system.

Each further and more complex stage is achieved by algorithmic modification reflecting on their biological organic parts. In that way genetic modification has been implemented according the Project. In the code system of each next species a maximal available modification is performed that to assure to it steady existence. Elements of algorithms of lowest organisms surely are still involved in the systems of highest modifications. Of course, their share progressively decreases by ascending through hierarchy, regardless; it is not impossible for them to be still building components and multiplied after each next modification.

Panteley Bahchevanov

SCIENCE AND RELIGION

By its contemporary development the science manifests an deep and steady entering in to the matter and micro world that gives a huge advantage of technologies and human's life. The space exploring as well revealed the material world at an unavailable before view point. These advantages of science and technologies give base to religion to look at a different way. From position of its discoveries science could declare, you see, all is material, development as well, we never and nowhere met something that to shows to opposite.

The hypothesis has already stated whether a material origin and development are probable. Now, would pay attention to that problem in the context of these contradictions between the visible and the confessed religions.

Talking about material base, consciously or not, but ever adding a spiritual component without which it should be inconceivable. The spirit and idea are also subjects of plenty of scientific works. But, how that abstract spirit without physical explanation to be incorporated in our material world and what should be its relation to the matter?

Let begin from Bible text. First, they are related to earth's habitants in ancient epoch in which our rational spirit to exist. We should say, the spirit of 100 years before is quite distinguished compared to ours, what to say about 3000 years before. The spirit of that epoch is really naïve towards the phenomena that surround us, the human's destiny and mode of life. Of course, there is and strong superstition. There are deviations from believe in one living Got and making a bow to idols. Then, all texts of Bible that seem today naïve we should relate to that epoch and level of reasoning. Would not comment the probability some of the texts to be rewritten during the next centuries. Seems, the texts regarding the human creation are in contradiction to reality and contemporary evolution theories.

But, there is a statement that corresponds to our hypothesis – all living including human are being created. We do not mean to reconcile religion and science; we simply follow our own train of reasoning, based on pragmatic physical references. Where are positioned our conclusions in the chain between religion and science is simply consequence, not an object. We should consider the science, despite its advantages, as well approaches to some level of dogmatism, keeping strictly to mathematical models, experiments and results. To enter in location of unknown and unexplored is risky but as well and a good reason to deny. When such serious physical phenomena as gravitation field are not rationally expounded this is not a reason to take position of closed eyes. It must be explained physically exactly as it is. Explanation living organism is being created by the material particles somehow does not satisfy contemporary people. The question is about a Supreme Reason and Logic. Let somebody repeats origin of living organism in some laboratory.

Finally, why the current material appears now, but did not yesterday. Because it is considerable people who talk and see one other at any point of world and expect tomorrow more and more advantages to be matured to accept all this information. Perhaps, it should be a message to the science as well.

It is probable religion to become more pragmatic, bringing to people exactly what they need to know and what could be understand from Holy Books – there is one Living Got. Our époque of technology and spiritual development is the time when people have to acknowledge that.

EVOLUTION SUMMARY

As at each part of that material, and here would accomplish with a brief abstract, where to try to order things to be clear at a glance.

WHY BIOLOGICAL?

- Biological structures could be synthesized and developed by controlling fields, integrated and specialized to do that
- Organisms based on biological matter are light, have low energy consumption, and could realize elastic joint, moving and many other complex functions
- Biologic matter permits building of extremely complex by design, physical principle

and functionality micro and macro structures, systems and organisms

- The built in that way micro and macro bodies, organs and systems allow realistic presence in to material world through optical, acoustic, temperature, mechanical and other physical sensors, chemical analyze
- Biological matter permits flexibility, reproduction and development
- By means of modification of the code systems of the controlling fields, mutation is being performed of the species
- Previously coded tolerances allow the organisms to adapt to environmental changes and to survive
- Biological evolution has given opportunities a living nature to be created that to harmonize to the features of planet
- Biological evolution has given opportunities to form human with features needed to obtain a reason

STAGES OF THE EVOLUTION

Note: The Hypothesis reviews evolution of matter, besides the evolution of organisms is a part of the whole. That classification is far from this to affect any contemporary theories and official statements.

ESSENCE OF THE EVOLUTION PROCESS

- Evolution process is planned and has object
- Each stage has been carefully considered and designed
- There are no missed links and stages
- The process is being performed precisely, methodically, besides each lowest stage is base of the next

BASIC FEATURES OF HUMAN TO OBTAIN A REASON

- Excellent moving functions
- Bipedal
- Function "Catch"
- Opportunity to access and presence in all environmental media
- Harmony with planet's size
- Exterior to correspond to mentality
- Destined to perform in material media activities related to energy consumption and apply of mechanical efforts that are unavailable for ethereal structures
- All activities of human are being controlled by low energy impulses of its controlling field
- In its activities human has full independence

***

The Hypothesis started from the idea the space in the Universe is being filled by a matter spread everywhere in all directions. The same idea came in minds of the physicists at the dawn of exploration and practical application of electromagnetic fields. Further that idea has been rejected because the apparatuses were unable to detect the ether and the experiments failed.

Let imagine the space in two variants, first as being empty and second as filled by a flexible net, of course in the context of an abstract considering only. It's a problem to consider in an empty space to action forces and tensions of strain, press, bend and twist. But when the flexible net fills that space, these tensions would be translated and spread to all the objects inside of it. Could imagine how the net should be deformed in all directions of the space.

Really these tensions correspond to gravitation forces and waves and lead to a static and dynamic distortion. In an empty space could establish an absolute coordinate system in each its point. If to a space object apply a coordinate system it will describe its relative motion towards the absolute. But, in an empty space the objects should have the freedom to move in all directions without regulations and mechanical restrictions. In real, they are mutually dependent one to other, distorting the space and to put an absolute coordinate system becomes available in determined points only.

We explained, a basic feature of the primary matter is the lack of energy, it consists of particles being absolute motionless. The time is ever related to a motion, therefore in that Media time does not exist. The concept of time is being introduced as a relative to the motion of the micro objects, particles or the space macro objects. In that way time is ever applied to the secondary matter, where a motion and energy exist. Similarly in the space does not exist an idea about up and down, this is related to our physiological senses on our planet.

Introducing anew the concept of ether led to unexpected consequences. It occurred a real opportunity to make a rational explanation of some problems wrapped in mist or accepted as a dogma. It led to disentangle a knot, of course, at that stage far from this to be in the focus of attention or acknowledgement of the science.

The hypothesis is far from a possibility to explain everything; it only slightly raises the curtain, giving an opportunity to peep outside. The logical chain started from the primary mater conducted us to the conclusion that giant masses of primary matter in the Universe could not be ever in a status of ideal balance. In some point of the space internal tensions should occur, as a consequence conditions the energy barrier to be overcome and a fusion to be initiated. In that way the matter is originated from the primary substance. Further the same train of reasoning led us to the conclusion the bodies are unable to attract one another but forces press them. And finally, we approached to the idea regarding the origin and development of life.

The hypothesis categorically claims, we are being designed and created precisely and methodically, and in the base of that scheme is our Creator. Introducing that rational and devoid of mystique concept we are but far from a further explanation, the knowledge in the location are strongly limited. The problem concerns the eternal dilemma what is primary and how the idea had been originated. We consider, the given in that context is enough and we do not have to ask for more. Regardless, the main idea of the hypothesis is acknowledgement of our Creator without mystique, explained only on a physical basis and argumentation.

The Hypothesis, a subject of that book surely should sound strangely especially for someone who is a skeptical disposed materialist. The Author relates with a great respect to every individual or social opinion and conception. But, what we could not to see and explore we should not reject easily because ever there is a possibility at some level to exist. What is logically more probable seems to be more available to exist.

The Author deeply hopes not to be thrown on the stake at the twenty first century for what he has afforded to write. Rather, or instead of that he appeals to considering and patience following the logical chains and argumentation before to reject anything as being not scientific and hence impossible. Let imagine a contemporary apparatus distantly operated. Should be able to operate it without coded signals and physical fields? Indeed, the living organisms have no distant operation being independent but without a complex code systems how should function?

Simultaneously with the entering of the current book to publishing a second one is being initiated where some of the same topics are reviewed but at a more popular way and in some changed context. It contents a non-standard concept regarding the social development as well. In addition to that would pay attention to gravitation field management, some ancient and contemporary world mysteries such as extraterrestrials in ancient times, geoglyphs and lines on earth's surface, crop circles and more.

www.ingramcontent.com/pod-product-compliance
Lightning Source LLC
Chambersburg PA
CBHW071419170526
45165CB00001B/335
* 9 7 8 1 4 1 9 6 7 0 2 3 7 *